D1236848

The Estuarine Ecosystem

TERTIARY LEVEL BIOLOGY

A series covering selected areas of biology at advanced undergraduate level. While designed specifically for course options at this level within Universities and Polytechnics, the series will be of great value to specialists and research workers in other fields who require knowledge of the essentials of a subject.

Recent titles in the series:

Mammal Ecology	Delany
Virology of Flowering Plants	Stevens
Evolutionary Principles	Calow
Saltmarsh Ecology	Long and Mason
Tropical Rain Forest Ecology	Mabberley
Avian Ecology	Perrins and Birkhead
The Lichen-Forming Fungi	Hawksworth and Hill
Social Behaviour in Mammals	Poole
Physiological Strategies in Avian Biology	Philips, Butler and Sharp
An Introduction to Coastal Ecology	Boaden and Seed
Microbial Energetics	Dawes
Molecule, Nerve and Embryo	Ribchester
Nitrogen Fixation in Plants	Dixon and Wheeler
Genetics of Microbes (2nd edn.)	Bainbridge
Seabird Ecology	Furness and Monaghan
The Biochemistry of Energy Utilization in Plants	Dennis
The Behavioural Ecology of Ants	Sudd and Franks
Anaerobic Bacteria	Holland, Knapp and Shoesmith
An Introduction to Marine Science (2nd edn.)	Meadows and Campbell
Seed Dormancy and Germination	Bradbeer
Plant Growth Regulators	Roberts and Hooley
Plant Molecular Biology (2nd edn.)	Grierson and Covey
Polar Ecology	Stonehouse

TERTIARY LEVEL BIOLOGY

The Estuarine Ecosystem

Second Edition

DONALD S. McLUSKY, B.Sc., Ph.D.

Senior Lecturer in Biology
University of Stirling

Blackie

Glasgow and London

Published in the USA by
Chapman and Hall
New York

Blackie and Son Limited,
Bishopbriggs, Glasgow G64 2NZ
7 Leicester Place, London WC2H 7BP

Published in the USA by
Chapman and Hall
a division of Routledge, Chapman and Hall, Inc.
29 West 35th Street, New York, NY 10001-2291

First published 1981

British Library Cataloguing in Publication Data

McLusky, Donald S. (Donald Samuel)
The estuarine ecosystem.—2nd ed.
1. Estuary ecosystems
I. Title II. Series
574.5'26365

ISBN 0-216-92671-8
ISBN 0-216-92672-6 pbk

Library of Congress Cataloging-in-Publication Data

McLusky, Donald Samuel
The estuarine ecosystem/Donald S. McLusky
p. cm.—(Tertiary level biology)
Bibliography: p.
Includes index
ISBN 0-412-02091-2 (Chapman and Hall).—ISBN 0-412-02101-3
(Chapman and Hall: pbk.)
1. Estuarine ecology. I. Title II. Series
QH541.6.E8M32 1990 89-15732
574.5'26365—dc20. CIP

Phototypesetting by Thomson Press (I) Ltd.
Printed in Great Britain by Bell & Bain (Glasgow) Ltd.

Preface

For the inhabitants of many of the world's major cities and towns, estuaries provide their nearest glimpse of a natural habitat; a habitat which, despite the attempts of man to pollute it or reclaim it, has remained a fascinating insight into a natural world where energy is transformed from sunlight into plant material, and then through the steps of a food chain is converted into a rich food supply for birds and fish. The biologist has become interested in estuaries as areas in which to study the responses of animals and plants to severe environmental gradients. Gradients of salinity for example, and the problems of living in turbid water or a muddy substrate, prevent most animal species from the adjacent sea or rivers from entering estuaries. In spite of these problems, life in estuaries can be very abundant because estuarine mud is a rich food supply which can support a large number of animals with a large total weight and a high annual production. Indeed estuaries have been claimed to be among the most productive natural habitats in the world.

When the first edition of this book appeared, biologists were beginning to realise that the estuarine ecosystem was an ideal habitat in which to observe the processes controlling biological productivity. In the intervening period, several more estuaries and their inhabitants have been studied intensively, and it is now possible to answer many of the questions posed by the earlier edition, and to pursue further the explanation of high productivity in estuaries and of energy utilisation at different trophic levels within the estuarine food web.

Users of the first edition were kind enough to welcome the framework of the book, which first outlined the estuarine environment and the physical and biological factors which are important within it. In this new edition, we also examine the responses of the animals and plants to these factors, consider the problems of life in estuaries and why so few species have adapted to estuaries, and then propose a food web for an estuary. Thereafter we shall examine each trophic level in the food web in turn, first the primary producers (plants and detritus), then the primary consumers (herbivores and detritivores) and finally the secondary consumers (carnivores). In this new edition more examples of the organisms in each trophic level will be discussed where new information has become available, and particular attention will be paid to new studies of the relationships between each trophic level.

In the period since the publication of the first edition of this book, a vast amount of new information on pollution in estuaries has accumulated. It has been widely recognised that although the world's seas are huge and may appear capable of receiving unlimited quantitites of man's waste, such waste is almost always discharged first into the confined waters of estuaries. Many international experts have stated that, whilst the open oceans may not be generally polluted, the coastal waters of the sea and especially the waters of estuaries are widely polluted. Thus in practice, marine pollution is often essentially estuarine pollution. To reflect this large impact of mankind on estuaries, and to consider how mankind may either destroy or enrich the estuarine ecosystem, two completely new chapters have been prepared in this edition. The first of these considers pollution in estuaries, and the diverse uses and abuses of the estuarine habitat by man, whilst the second new chapter considers the ways in which estuarine management can either monitor, control or prevent pollution or destruction of the estuarine ecosystem.

This new edition therefore retains the concept of the study of the ecosystem as the basis for our understanding of the natural world, and shows that estuaries are ideal habitats for such studies. The new content and chapters reflect our attempts to recognise both the problems of pollution in estuaries and the solutions which estuarine management can offer, as estuarine ecosystems come under increasing pressure from a wide range of demands made by an increasing world population. The reading list has been completely updated to reflect recent progress in estuarine science.

May I thank the following authors and publishers for allowing me to use copies of their figures: Peter Meadows and Jan Campbell (Blackie; 2.1, 2.2), Kai Olsen (*Ophelia*; 2.4, 3.1, 3.3), Reni Laane (Springer-Verlag; 2.8), Colin Moore (Royal Society of Edinburgh; 3.8, 3.9), Sven Ankar and Ragnar Elmgren (*Askö Contributions*; 3.11), Colin Taylor (Royal Society of Edinburgh; 3.14, 3.15), Peter Burkhill (Plymouth Marine Laboratory; 3.16), Tom Pearson and Rutger Rosenberg (*Oceanography and Marine Biology, Annual Review*; 5.5, 5.6). Other figures were prepared by the author. Sources of information which are not mentioned in the text are listed in the reading lists.

It is a pleasure to thank Khlayre Mullin and June Watson for their help with typing, and Lewis Taylor for his photographic assistance. I would like to thank my colleagues for their valuable comments, especially Mike Elliott who contributed so much to Chapter 6, and Tony Berry and David Bryant. This book is dedicated to my patient family, Ruth, James, Sarah and Becky.

DONALD S. McLUSKY

Contents

CHAPTER ONE
THE ESTUARINE ENVIRONMENT

1.1 Introduction

Estuaries have for long been important to mankind, either as places of navigation, or as locations on their banks for towns and cities. Nowadays they are under pressure, either as repositories for the effluent of industrial processes and domestic waste, or as prime sites for reclamation to create land for industry or agriculture. Against this background the biologist has been attracted to other functions of estuaries: vital feeding areas for many species of birds, especially waders and wildfowl, the locations of coastal fisheries, or as fascinating areas which present challenges to our understanding of how animals and plants adapt to their environment. The estuarine environment is characterised by having a constantly changing mixture of salt and fresh water, and by being dominated by fine sedimentary material carried into the estuary from the sea and from rivers which accumulates in the estuary to form mudflats. The mixtures of salt and fresh water present challenges to the physiology of the animals which few are able to adapt to. The mudflats present areas which are rich in food, but are low in oxygen or even anoxic.

In this book I shall examine estuaries by trying to understand the interactions and feeding relationships which make up the estuarine food-web. Estuaries have been claimed to be the most productive natural habitats in the world, and I shall attempt to explain why they are so productive, and how the energy produced is utilised by succeeding trophic levels. Before we can examine the life of an estuary we must first examine the physical and chemical features which mould the estuarine environment.

The most useful definition of an estuary has been given by Pritchard (1967) as: 'an estuary is a semi-enclosed coastal body of water, which has a free connection with the open sea, and within which sea water is measurably diluted with fresh water derived from land drainage'. This definition of estuaries in particular excludes coastal lagoons or brackish seas from our consideration. Coastal lagoons for example do not usually have a free connection with the open sea and may only be inundated with sea water at irregular intervals. Brackish seas such as the Caspian Sea may have salinities comparable to some parts of estuaries, but they do not experience the regular fluctuations of salinity due to tidal effects. The definition of 'semi-enclosed' serves to exclude coastal marine bays, and the definition of 'fresh water derived from land drainage' serves to exclude saline lakes with fresh water from rainfall only. An estuary is thus emphasised as a dynamic ecosystem having a free connection with the open sea through which sea water enters normally according to the twice-daily rhythm of the tides. The sea water that enters the estuary is then measurably diluted with fresh water flowing into the estuary from rivers. The patterns of dilution of the sea water by fresh water varies from estuary to estuary depending on the volume of fresh water, the range of tidal amplitude and the extent of evaporation from the water within the estuary.

The salinity of sea water is approximately 35‰ NaCl, tending to be lower (33‰) in coastal seas and higher (37‰) in tropical waters. The salinity of fresh water is always less than 0.5‰. Thus the salinity of estuarine waters is between 0.5 and 35‰. This range is generally termed brackish, as distinct from marine or fresh waters. Salinity is a measure of the salt content of the water, and is expressed as total concentration of salts in grams contained in one kilogram of sea water. The salts are principally sodium and chloride ions, supplemented by potassium, calcium, magnesium, and sulphate ions, plus minute or trace amounts of many other ions. Whereas marine and fresh waters are characterised by stable salinities, estuarine water is extremely variable in its salinity.

The pattern of salinity distribution within estuaries may be used as a basis for the classification of estuaries. Three main types of estuaries can be recognised in the world, namely positive, negative or neutral estuaries. In positive estuaries the evaporation from the surface of the estuary is less than the volume of fresh water entering the estuary from rivers and land drainage. In such a positive estuary (Figure 1.1) the outgoing fresh water floats on top of the saline water which has entered the estuary from the sea, and water gradually mixes vertically from the bottom to the top. This type of estuary, which is the most typical in the temperate parts of the world, is thus characterised by incoming salt water on the bottom, with gradual vertical mixing leading to an outgoing stream of fresher surface water.

In negative (or inverse) estuaries the opposite situation exists, with evaporation from the surface exceeding the freshwater run-off entering the estuary. This type of estuary is mostly found in the tropics, for example the

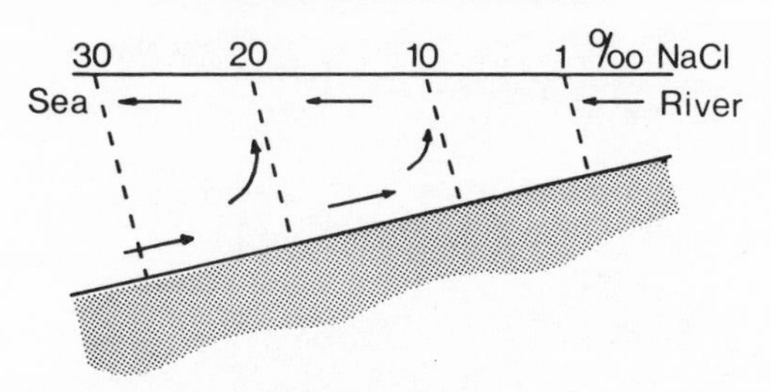

Figure 1.1 Positive estuary. Freshwater run-off is greater than evaporation. The arrows show the pattern of circulation with denser marine water entering the estuary along the bottom and then gradually mixing vertically with the outgoing surface stream of fresher water. Salinity is expressed as ‰ NaCl.

Laguna Madre, Texas, or Gulf St. Vincent, South Australia, although it can also occur in temperate regions where the freshwater input is limited such as the Isefjord in Denmark. In negative estuaries evaporation causes the surface salinity to increase. This saltier surface water is then denser than the water underneath and thus sinks. The circulation pattern is thus opposite to that of a positive estuary, because in a negative estuary the sea and fresh water both enter the estuary on the surface, whence after evaporation and sinking they leave the estuary as an outgoing bottom current (Figure 1.2). Occasionally the fresh water input to the estuary exactly equals the evaporation and in such situations a static salinity regime occurs (Figure 1.3). Such an estuary is termed a neutral estuary, but they are rare as evaporation and fresh water inflow are almost never equal.

In this book the emphasis throughout will be on the commoner positive type of estuary. In general, we can thus already see that estuaries are a

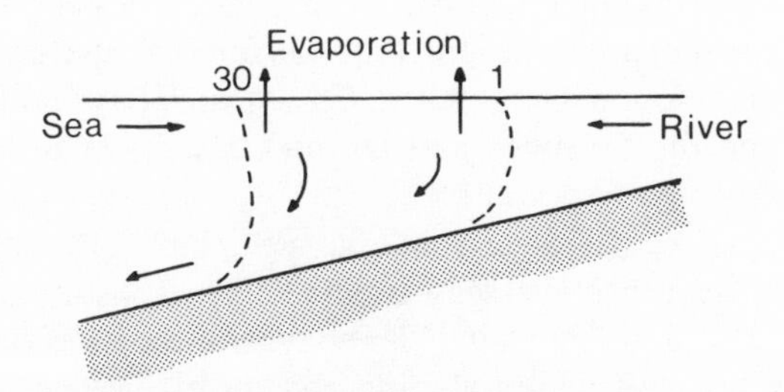

Figure 1.2 Negative estuary. Freshwater run-off is less than evaporation. The arrows show the pattern of circulation with both sea water and fresh water entering the estuary at the surface. Within the estuary, evaporation produces denser high-salinity water which sinks to the bottom. Salinity is expressed as ‰ NaCl.

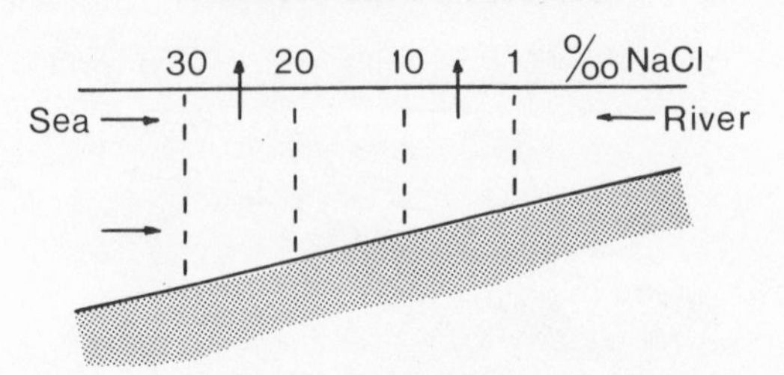

Figure 1.3 Neutral estuary. Freshwater run-off is equal to evaporation. This situation is intermediate, and possibly transitory, between a positive estuary (Figure 1.1) and a negative estuary (Figure 1.2). Salinity is expressed as ‰ NaCl.

habitat intermediate between the sea, the land and fresh waters: a habitat which is a complex dynamic mixture of transitional situations and which is almost never static. Within this dynamic environment physical and chemical factors show marked variations. Many of these factors are closely linked to the salinity distribution patterns mentioned above, such as the strength of currents, tidal amplitude, wave strength, and deposition of sediments as well as temperature, oxygen and the supply of nutrients.

Just as the physical and chemical factors within estuaries are in a continual state of change, so even the topography of estuaries is continually changing. Most estuaries have developed on the low-lying ground which forms the coastal plains around land masses. Here estuaries typically develop in river valleys which become drowned by the sea. In the drowned valleys extensive deposition of sediments carried into the estuaries has produced the estuaries that we see today. Drowned river valleys, also known as coastal plain estuaries or rias, have been formed since the last Ice Age, that is within the last 15 000 years. The height of the sea in relation to the land is continually changing, and whilst the mean sea level has increased as ice caps have melted, so in some areas the land has also subsided and in other areas the land that was formerly glaciated has sprung up as the weight of ice on it has melted. Chesapeake Bay and Delaware Bay in the USA, and the Cornwall and Devon estuaries in Britain, are good examples of drowned river valleys.

Other types of estuaries are lagoon-type, bar-built estuaries where shallow basins have become semi-isolated from coastal seas by barrier beaches composed mainly of sand. These type of estuaries occur on stable continental edges with gently sloping coastal plains and active coastal deposition, for example Barnegat Bay, New Jersey, Laguna Madre, Texas, Albufeira, Portugal and in parts of Australia. Fjord-type estuaries occur where valleys have been deeply eroded by glaciation, such as the coasts of Norway, Western Scotland, Alaska, and New Zealand and typically have deep inner basins linked to the sea by shallow entrance sills. A final estuarine type are those produced tectonically, as a result of land movements associated with faulting and volcanism. San Francisco Bay is the best known example of a tectonically produced estuary.

As siltation of estuaries occurs and salt marshes extend and consolidate the land, so the waters of the estuary may be pushed seawards. Small changes in sea level due to long-term climatic change could either drown or expose many of our present estuaries, although of course new ones may be formed elsewhere. The various types of estuary, however formed, are all characterised by the variability of their environmental parameters, most notably in their water circulation and their sediments.

1.2 Estuarine circulation

Estuaries may thus be classified as positive, neutral or negative depending on their salinity regime and the extent of evaporation. Depending on the tidal amplitude and volume of freshwater flow four main types of positive estuaries can be recognised (Dyer, 1973, 1979):

1. Highly stratified — example: Mississippi (USA), N. Esk (UK)
2. Fjords — example: Norwegian fjords, West of Scotland lochs
3. Partially mixed — example: James River (USA), Mersey, Thames, Forth (UK), Elbe (FRG)
4. Homogeneous — example: Delaware, Raritan (USA), Solway Firth (UK), Mandovi-Zuari (India)

In the first of these, the highly stratified or salt wedge, the fresh water flows seawards on the surface of the inflowing salt water. At the interface between the salt and fresh water, entrainment (mixing) occurs and salt water is mixed into the outflowing fresh water (Figure 1.4). For this type of estuary a large river flow in relation to tidal flow is needed and continuous downstream flow of surface water may occur despite the ebb and flood of the salty tidal water beneath it. The second type, the fjord type, is basically similar to the highly stratified, except that due to a sill (a shallow lip) at the mouth of fjords the inflow of tidal water is more restricted. Again a continuous downstream flow of fresh water at the surface occurs, but the renewal of tidal water may only occur seasonally and non-renewal of water may lead to anoxic conditions in the deepest parts of the fjord (Figure 1.5).

Given a situation with the tidal inflow greater than, or similar to, the fresh water inflow a partially mixed estuary develops. In such an estuary

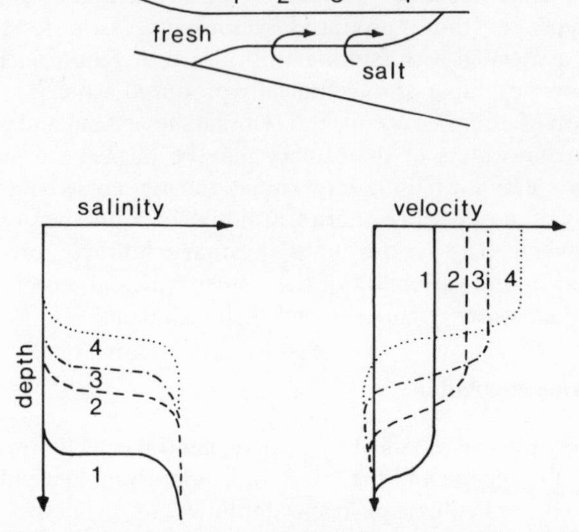

Figure 1.4 Salinity and velocity profiles in a salt wedge estuary. The salinity and velocity profiles for 4 positions in the estuary are shown, with arbitrary axes. (After Dyer, 1973.)

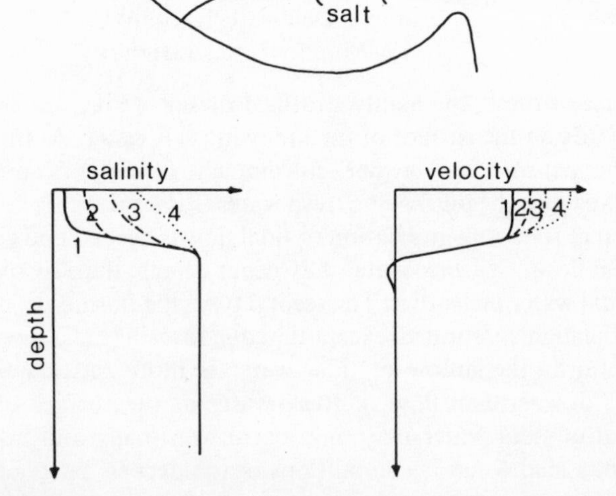

Figure 1.5 Salinity and velocity profiles in a fjordic estuary. The salinity and velocity profiles for 4 positions in the estuary are shown, arbitrary axes. (After Dyer, 1973.)

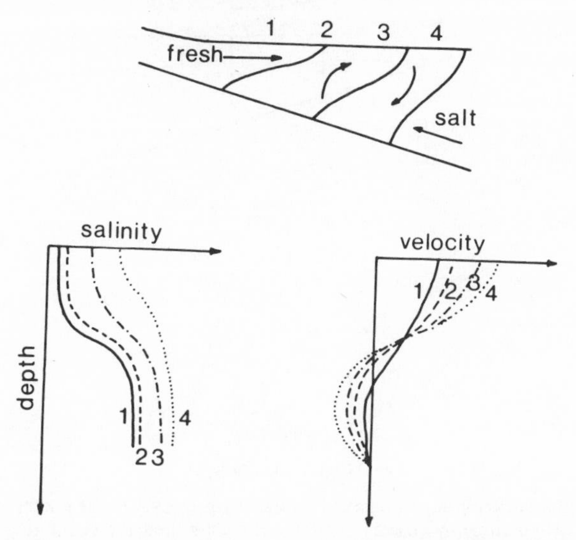

Figure 1.6 Salinity and velocity profiles in a partially mixed estuary. The salinity and velocity profiles for 4 positions in the estuary are shown, with arbitrary axes. (After Dyer, 1973.)

there is a continuous mixing between the sea and fresh water. Surface waters will be less saline than the bottom waters at any given point in the estuary, but unlike the highly stratified estuary undiluted fresh water will only be found near the head of the estuary. Mixing of water from the predominantly inflowing bottom to the mainly outflowing surface will occur throughout the estuary (Figure 1.6). The pattern of mixing may be less clear at the margins of the estuary, and due to the Coriolis force the sea water will dominate on the left-hand side (looking downstream) in the northern hemisphere, and the outgoing fresher water will dominate the right-hand side. In the southern hemisphere, the dominance pattern is reversed.

When the estuary is very wide the Coriolis force will cause a horizontal separation of the flow, with outgoing flow on the right-hand side in the northern hemisphere, and ingoing flow on the left-hand side. Thus the circulation in such a homogeneous estuary will be across the estuary from left to right, rather than vertically as in the other types (Figure 1.7).

The range of tidal amplitude varies in a constant pattern in the seas of the world and the tides within estuaries are due to the tidal wave at their mouth. In the waters around Denmark and Sweden, for example, the tidal amplitude is very small and the many estuaries of Denmark are not subject

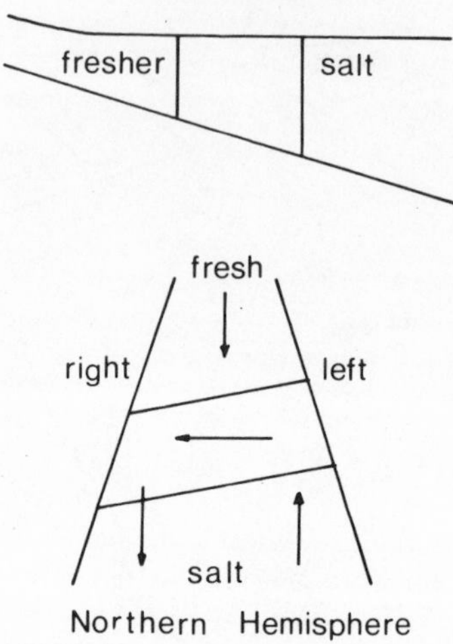

Figure 1.7 Salinity distribution in a large homogeneous estuary in the northern hemisphere. Due to the Coriolis force, horizontal separation occurs, and the circulation is across the estuary rather than vertically. (Modified after Dyer, 1973.)

to strong tidal currents and do not develop large intertidal areas. At the other extreme the estuaries of the Canadian Bay of Fundy region experience maximal tidal amplitude producing strong tidal currents as well as large intertidal areas. Tidal amplitude varies not only from place to place but it also varies at any particular locality according to the neap and spring tide cycle, as the range of tides fluctuates from a maximum rise and fall at spring tides to a lesser rise at neap tides. The full cycle from spring to neap tides occurs each lunar month (28 days). In all estuaries the strength of tidal and river currents is dissipated as the main inflow currents collide and mix with each other, and compared to both the sea and rivers, estuaries are quiet places. Due to the shelter of the land surrounding the estuary, wind-driven waves are much smaller than in the sea. This absence of large waves, coupled with the mixing of currents is, as we shall see, vital for the deposition of estuarine sediments.

The river flow, tidal range and sediment distribution in estuaries are continually changing and consequently estuaries may never really be 'steady-state' systems, and may be tending towards a balance that they never achieve. With increased river flow, or during neap tides, the extent of tidal intrusion of sea water diminishes, whilst with spring tides or decreased river flow the tidal intrusion increases. The concept of 'flushing time' is one attempt to relate the fresh water flow to the tidal range, and the volume of the estuary. The flushing time (T) is the time required to replace the existing fresh water in the estuary by the river discharge, and is calculated by dividing the total volume of river water accumulated in the estuary (Q) by the river flow (R) (Dyer, 1973) thus: $T = Q/R$.

In the Forth estuary for example, the flushing time under mean river flow conditions is 12 days, but in summer with reduced river flow it may be up to 10 weeks, and following severe rainfall it may be down to 6 days. These values need, of course, to be calculated individually for each estuary. Their calculation can be of great value in predicting the impact of polluting discharges on estuaries, as an estuary with a short flushing time is generally better mixed and better able to accept effluents.

Within particular estuaries the pattern of circulation may also vary throughout the year in response to meteorological events. The Potomac estuary (USA) is a partially mixed estuary for 43% of the year, with surface outflow and deep inflow, but the reverse with surface inflow and deep outflow also occurs for 20% of the year, and four other patterns occurred for the rest of the time in response to climatic events.

Techniques for the measurement of estuarine currents, tides and salinity are given by Dyer (1979) and Morris (1983). The salinity at any particular point of an estuary depends on the relationship between the volume of tidal sea water and the volume of fresh water entering the estuary, as well as the tidal amplitude, the topography of the estuary and the climate of the locality. The various hydrographic regimes of estuaries have been presented earlier, where it was seen that in positive estuaries the maximum salinity was always at the bottom of the water column. For the animals and plants living within estuaries, salinity presents a challenge to their physiological processes which will be discussed in section 1.6. At the moment, we can note that it is possible to divide estuaries into zones on the basis of their salinity. Following a symposium in Venice an agreed scheme of the classification of estuaries and brackish waters has been devised, and is presented in Table 1.1.

Whilst the salinity of the water bodies of estuaries is important for fish and planktonic organisms living in the water column, it is of less direct importance for the majority of estuarine animals which live buried within the muddy deposits. Far more important for these benthic animals is the interstitial salinity, which is the salinity of the water between the mud particles. It has been consistently shown that the interstitial salinity varies much less than the salinity of the overlying water, due to the slow rate of interchange between them. On the intertidal mudflats, where the most abundant populations of estuarine animals are often found, the interstitial

Table 1.1 Venice system for the classification of brackish waters. From 'Venice System' (1959).

Zone	*Salinity (‰ NaCl)*
Hyperhaline	> ± 40
Euhaline	± 40 – ± 30
Mixohaline	(± 40) ± 30 – ± 0.5
Mixo-euhaline	> ± 30 but < adjacent sea
-polyhaline	± 30 – ± 18
-mesohaline	± 18 – ± 5
-oligohaline	± 5 – ± 0.5
Limnetic (fresh water)	< ± 0.5

salinity matches that of the high-tide salinity when it covers the mudflats, which may be considerably higher than the salinity of the estuarine water at low tide at the same location. Because of this phenomenon it is usually possible for marine animals living buried in the sediment to penetrate further into estuaries, than for marine animals living planktonically.

1.3 Estuarine sediments

Fine sedimentary deposits, or muds, are a most characteristic feature of estuaries, and indeed the estuarine ecosystem has been defined by Hedgpeth (1967) as 'a mixing region between sea and inland water of such shape and depth that the net resident time of suspended (sedimentary) materials exceeds the flushing'. Sedimentary material is transported into the estuary from rivers or the sea, or is washed in from the land surrounding the estuary. In most North European and North American estuaries the main source of sedimentary material is the sea, and the material is carried into the estuary either as suspended sediment flux or as bedload transported in the bottom inflowing currents that characterise salt wedges. In the Tay estuary, Scotland, UK, for example, 70% of the sediments accumulating on the tidal flats are of marine origin. However in some French estuaries, where fine material is available on the banks of the estuary, these banks are the main source of estuarine sediments. In the estuaries of the Loire (France), Vigo (Spain), Apalachicola (USA) and Yellow River (China), rivers carrying large quantities of clay are the main source of estuarine sediments.

Whatever the source of the sediments the deposition of it within the estuary is controlled by the speed of the currents and the particle size of the

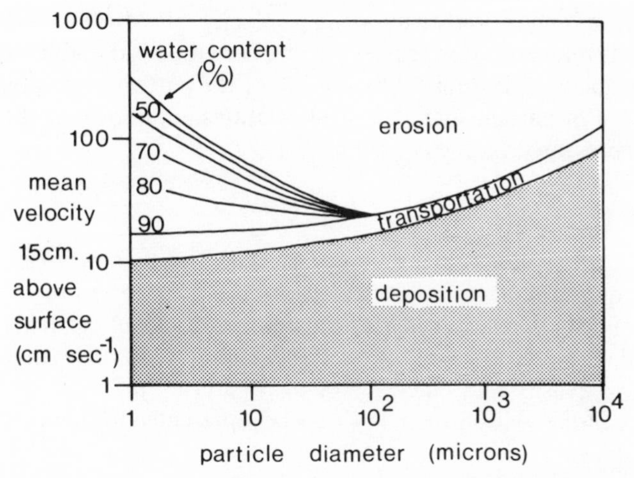

Figure 1.8 Erosion, transportation and deposition velocities for different grain particle sizes. The diagram also illustrates the effect of the water content of the sediment on the degree of consolidation which in turn modifies the erosion velocities. (After Postma, 1967.)

sediments. The relationship between current speed and the erosion, transportation and deposition of sediments is shown in Figure 1.8. It can be seen that for a sediment with pebbles 10^4 μm (1 cm) diameter erosion of the sediment will take place at current speeds of over 150 cm s^{-1}. At current speeds between 150 and 90 cm s^{-1} the pebbles will be transported by the current, but will be deposited at current speeds less than 90 cm s^{-1}. Similarly for a fine sand of 10^2 μm (0.1 mm) diameter erosion will occur at speeds greater than 30 cm s^{-1}, and deposition will occur at speeds less than 15 cm s^{-1}. For silts and clays with particles of 1–10 μm diameter a similar relationship occurs, except that consolidation of the sediment means that for erosion to occur faster current speeds are needed as the consolidated sediment behaves as if it were composed of larger size particles. The consequence of these relationships for estuaries is that in the fast-flowing rivers and strong tidal currents at either end of the estuary all sizes of sedimentary particles may be eroded and transported. As the currents start to slacken within the estuary so the coarser pebbles and sands will be the first to be deposited, and the finer silts and clays will remain in suspension. In the calmer middle and upper reaches of an estuary, where the river and tidal currents meet, and especially in the slack water at high tide overlying the intertidal areas, only then will the currents be slow enough for mud to be deposited.

The rate of deposition, or the settling velocity, of sediments is related to the particle diameter. For coarser particles (medium sand and larger) the settling velocity is determined by the size of the particle and varies as the square root of particle diameter. This relationship, known as the impact law, is represented by the equation:

$$V = 33\sqrt{d}$$

where V is the settling velocity (cm s^{-1}) and d is the diameter of a spherical grain of quartz (cm). For finer particles (silts, clays and fine sands) the settling velocity is determined by the viscous resistance of the fluid in which the particle is settling and Stokes' law applies. Particle shape, concentration, density and efficiency of dispersion all influence the precise values for Stokes' law. Stokes' law may be represented for quartz particles in water at 16°C as:

$$V = 8100d^2$$

(V, d as above).

The results of the impact and Stokes' law as settling velocity of various sizes of sediments are presented in Table 1.2, from which it can be seen that sands and coarser materials settle rapidly in water, and any sediment coarser than 15 μm will settle within one tidal cycle. However for finer sediments, settling velocities are much slower and clay and silt particles less than 4 μm diameter will certainly be unable to fall and settle within one tidal

Table 1.2 Settling velocities. (After King, 1975.)

Material	*Median diameter (μm)*	*Settling velocity (m day^{-1})*
Fine sand	250–125	1040
Very fine sand	125–62	301
Silt	31.2	75.2
Silt	15.6	18.8
Silt	7.8	4.7
Silt	3.9	1.2
Clay	1.95	0.3
Clay	0.98	0.074
Clay	0.49	0.018
Clay	0.25	0.004
Clay	0.12	0.001

cycle. Consequently the waters of estuaries tend to be very turbid as the silt and clay particles in suspension are carried about the estuary until they eventually settle to form the mudflats which are so characteristic of estuaries. The speed of settlement can be slightly faster than portrayed in Table 1.2, as 'salt flocculation' can occur. This process is caused by the clay particles in salt water tending to adhere to each other, and as they form larger particles so they will tend to fall faster, for example particles of 1.5 μm in the deflocculated state form flocculated particles which behave as if they were 7 μm in diameter. Sea salts have two roles to play in the aggregation of fine particles; firstly, even a few parts per thousand of salt is sufficient to make suspended particles cohesive, and secondly, the collisions between particles prior to aggregation occur most frequently in the density gradients produced by the mixing of salt and fresh waters in estuaries.

The middle and upper reaches of estuaries are thus characterised by very turbid water with poor light penetration. Within many estuaries the suspended matter in the estuaries develop so-called 'turbidity maxima'. The presence and magnitude of turbidity maxima are controlled by a number of factors, including the amount of suspended matter in the river or sea water, the estuarine circulation, and the settling velocity of the available material. As Figure 1.9 shows for the Ems estuary the turbidity maxima is located at the meeting point of the river and tidal currents. The sedimentary material in the central and upper reaches of the estuary is continually added to by river discharge and tidal inflow, and remains largely in suspension, before finally settling. The position of the turbidity maximum is generally determined by the contribution of suspended sediment from the seaward end of the estuary. The inward flow of marine water and its sediment moves along the bottom of the estuary until it reaches a stagnation point where inward flow ceases. At this stagnation point the sea water and its sediment rise to mix with the surface fresher water. The position of this stagnation point, and the consequent turbidity maximum, also varies with the strength of river flow, moving upstream with low river flow, and downstream with high river flow.

In many estuaries the maximum concentration of suspended sediment occurs at low tide, as the ebbing tide washes sediment off the intertidal areas and allows the sediment in suspension to remain in the low water channel. As the tide rises, the concentration of suspended load is reduced as the flooding tide increases the volume of water in the estuary, and the sediments are carried over the intertidal areas, where settlement may occur at high tide (Figure 1.10). In northern latitudes, ice break-up in the spring is the most significant factor affecting muddy intertidal sedimentation.

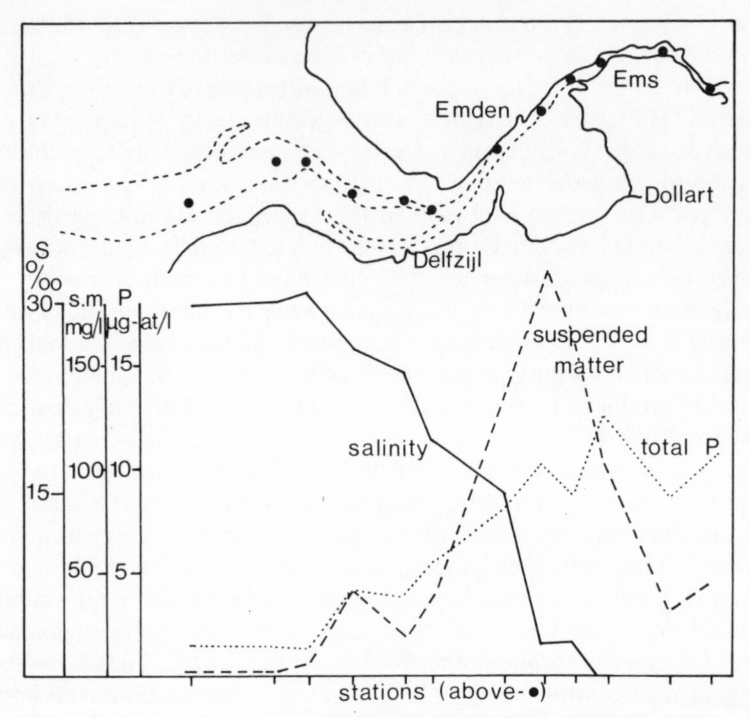

Figure 1.9 Turbidity maximum in the Ems-Dollard estuary, on the border between The Netherlands and the Federal Republic of Germany. The suspended matter is expressed as mg l^{-1}, salinity as ‰ NaCl, and total phosphorus as μg-at l^{-1}. Observations were taken at the surface of the estuary at the stations indicated on the map. (After Postma, 1967.)

During summer, organic processes dominate as organisms feed on, pelletise and bioturbate the sediment, whilst plants may stabilise the sediment. In autumn (fall) seasonal storms increase, and from then into winter most sediment erosion and transportation will occur (Anderson, 1983).

Along with the sediments being carried into estuaries are usually carried particles of organic debris derived from the excretion, death and decay of plants and animals. Once the dissolved and particulate organic matter reaches estuaries from fresh and salt water it tends to remain there as it is deposited and incorporated into the estuarine ecosystem along with fine inorganic matter. The organic matter in all aquatic ecosystems is conventionally divided ino two fractions by filtration through filters with an average pore diameter of 0.5 μm. The fraction passing through the filter is

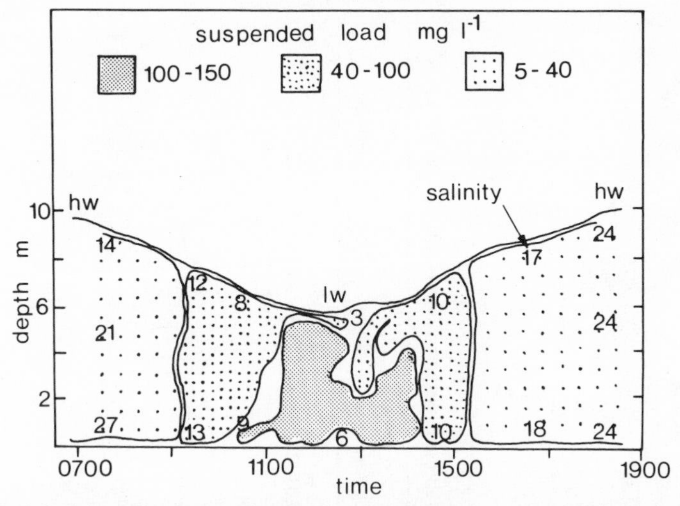

Figure 1.10 Salinity and suspended load variations in the Tay estuary, Scotland, during one tidal cycle of high water to low water and back to high water. Salinity values, as ‰ NaCl, are superimposed on the suspended load values (as mg l^{-1}). (After Sholkovitz, 1979.)

termed dissolved organic matter (DOM) even though it will contain very fine particulate matter in addition to truly dissolved matter. The fraction retained by the filter is termed particulate organic matter (POM). Typical concentrations of DOM and POM as organic carbon in rivers, estuaries and the sea are presented in Table 1.3, which shows estuarine water as intermediate between river water and the sea.

The organic matter within estuaries consists of material resulting from the excretion and decomposition of estuarine animals and plants, supplemented by fragments and dissolved organic matter carried into the estuary. Within the estuary the organic matter may be cycled and transformed, as shown in Figure 1.11. The material lost as exudation from plants, and as excretion from animals contributes to the dissolved organic material in the estuary, whereas the detritus produced from the death of organisms is primarily in the form of particulate matter. Some of the main processes involved in the flow of organic matter through the early stages of an estuarine ecosystem are shown in Figure 1.12.

The sedimentation of both inorganic and organic suspended material leads to the development of mudflats and other areas of deposition within estuaries. Attempts have been made to summate the sources of suspended material, and calculate the rate of deposition of sediments. For example,

Table 1.3 Concentrations of organic carbon in natural waters, (figures in brackets represent extreme values). (after Head, 1976.)

Concentration ($mg\,l^{-1}$)			*Coastal*	*Open sea*		
of organic carbon	*River*	*Estuary*	*sea*	*Surface*	*Deep*	*Sewage*
Dissolved (DOC)	10–20 (50)	1–5 (20)	1–5 (20)	1–1.5	0.5–0.8	100
Particulate (POC)	5–10	0.5–5	0.1–1.0	0.01–1.0	0.003–0.01	200
Total	15–30 (60)	1–10 (25)	1–6 (21)	1–2.5	0.5–0.8	300

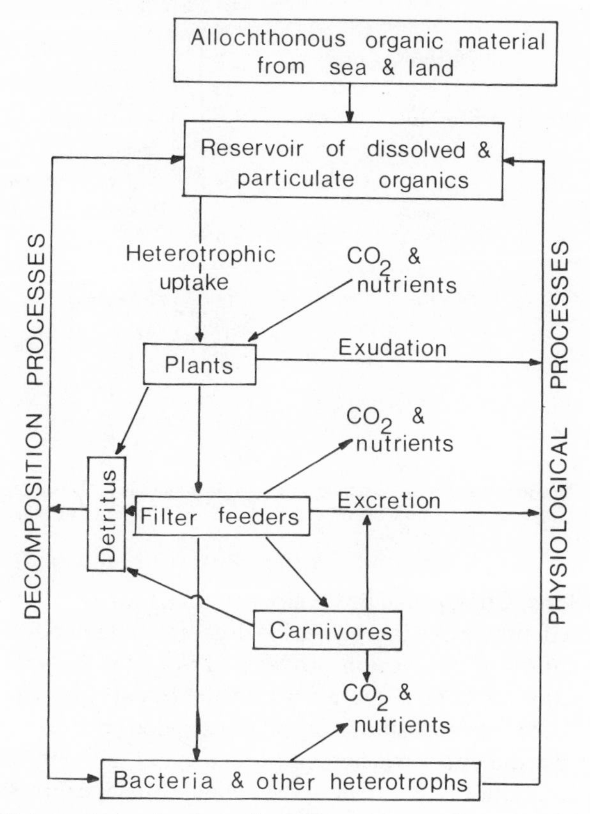

Figure 1.11 Major pathways for the cycling of organic matter in an estuary. (After Head, 1976.)

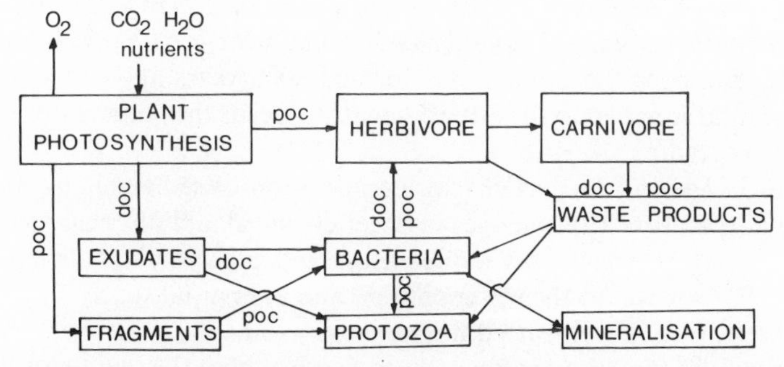

Figure 1.12 The flow of organic matter through an estuarine food web. doc = dissolved organic carbon, poc = particulate organic carbon. (After Head, 1976.)

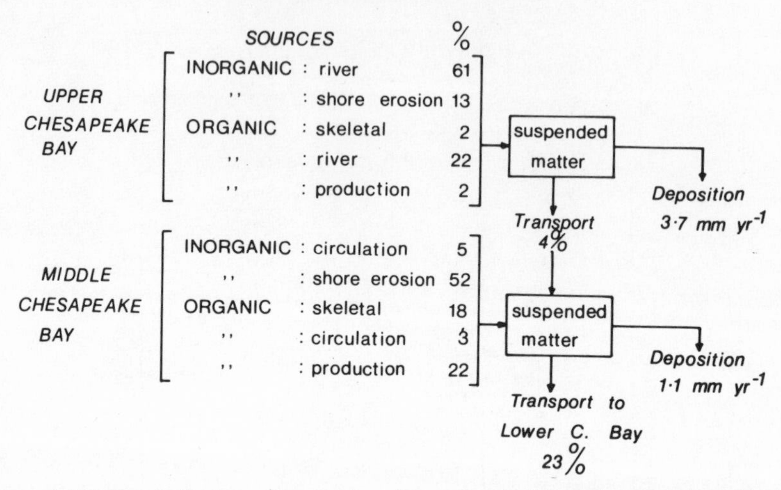

Figure 1.13 The contributions of inorganic and organic materials to the suspended sediment load of Upper and Middle Chesapeake Bay, USA. (After data in Biggs, 1970.)

within the upper Chesapeake Bay regions (USA), the river supplies 83% of the suspended matter (3/4 inorganic, 1/4 organic). In the middle reaches of the same estuary, shore erosion provides 52% of the suspended matter, with production of organic matter within the estuary providing another 22%. The rate of deposition of all suspended matter amounts to 3.7 mm yr^{-1} in the upper reaches, and 1.1 mm yr^{-1} in the middle reaches, as shown in Figure 1.13. Within the whole Chesapeake Bay system, including the lower reaches, deposition of sediment is occurring at a rate of 0.8 mm yr^{-1}, with 56% of the sediment supplied by rivers, 31% by shore erosion and 12% from tidal inflow. Elsewhere in the USA, in the Patuxent River estuary, sediment is collecting at a rate of 3.7 cm yr^{-1}, and in the large estuarine system of Long Island Sound, near New York, it has been shown that since the estuary was formed 8000 years ago the volume of marine mud supplied to the estuary greatly exceeds the volume supplied by the rivers feeding the area.

The changes in the particle size composition of sediments in estuaries noted above also cause changes in other chemical and physical properties of the sediment, which may influence the animals and plants living there. These changes are: to the water content and interstitial space, with higher water retention in well-sorted fine-grained sediments; to temperature and salinity where they change much more slowly within the sediments than in overlying water or air; and to organic content and oxygen content, where fine-grained sediments are associated with greater organic content, and due to the biological processes within them, with lower oxygen content. The change from an aerated surface sediment layer to deeper anoxic layers takes place closer to the surface in fine-grained muds than in coarse-grained sands. This change which may be approximated to by measuring the redox potential (E_h) of the sediment (see Pearson and Stanley, 1979), serves to limit the macrofauna in estuarine sediments to species which can form burrows or have other mechanisms to obtain their oxygen from the overlying water. The general relationships between benthic fauna and the nature of the sediments is reviewed by Rhoads (1974) and Gray (1981).

1.4 Other physico-chemical factors

It is tempting to regard estuarine water simply as diluted sea water. However there are dangers in this approach, as evidenced by Figure 1.14. It can be seen that the concentration of bicarbonate ion drops only slightly from the sea to fresh water. The concentration of other ions deviates markedly from the hypothetical dilution line at low salinities (i.e. at less than

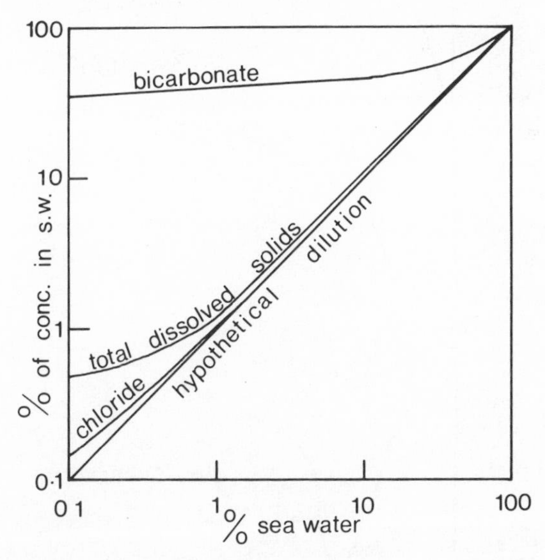

Figure 1.14 The effect of the dilution of sea water by river water on the concentrations of bicarbonate and chloride ions, and the concentration of total dissolved solids. Concentrations expressed as percentage of the concentration in sea water. Note the logarithmic scales. (After Phillips, 1972.)

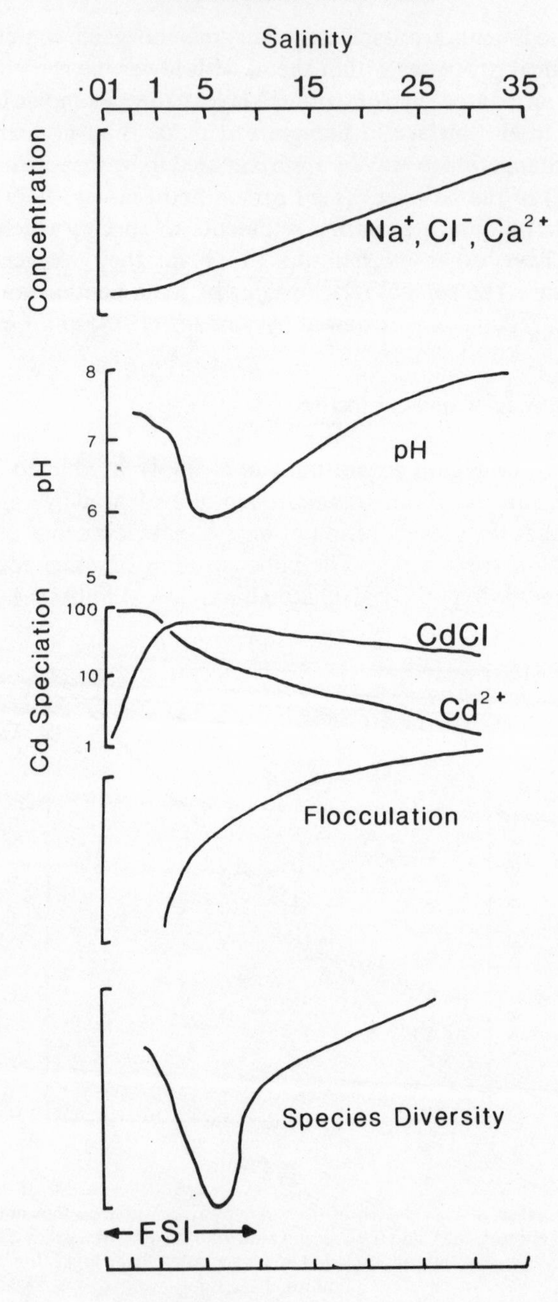
Salinity
·01
1
5
15
25
35
Concentration
Na⁺, Cl⁻, Ca²⁺
pH
8
7
6
5
pH
Cd Speciation
100
10
1
CdCl
Cd²⁺
Flocculation
Species Diversity
FSI

10‰), and certain ions such as phosphate, nitrate or silicate may even be more abundant in river water than sea water, and thus will show a decrease in concentration from rivers to the sea. Dissolved organic matter may also be higher in river water than the sea, and the concentrations of most trace metals are similar in river and sea water and therefore tend to be relatively constant within estuarine water. The distribution of trace metals, such as iron, manganese, cobalt, nickel, copper, zinc and cadmium in estuaries is controlled mainly by the distribution and transport of suspended particulate matter, especially the position of the turbidity maximum.

Most of the biogeochemical processes involved in the meeting of fresh water and sea water in estuaries occur at very low salinities (< 1‰), and Morris *et al.* (1978), along with Mantoura (1987) have identified in the Tamar estuary, a region known as the freshwater–seawater interface (FSI) as an important site for chemical and biological reactions. The FSI has also been termed the FBI (freshwater–brackishwater interface). The chemical and other changes occurring at the FSI are summarised in Figure 1.15, where it should be noted that the salinity axis is on a logarithmic scale, in order to expand the scale for lower salinities. From this figure it can be noted that changes in the compositional ratios of the major ions such as sodium, and chloride, resulting from mixing river water with sea water are complete within the first 1‰. These changes are accompanied by dramatic changes in the speciation of metals such as cadmium, with a drastic drop in the proportion of free cadmium ions, and a surge in the concentration of cadmium chloride complexes. The hydrogen ion activity (pH) is at a minimum at 0.5–3‰, and the drop in pH controls many aquatic reactions, such as dissolution of metal oxides. Thus the addition of sea salt to river water, with its low ionic strength, results in non-linear chemical perturbations that are amplified in the low salinity waters of the FSI. Recent studies on the Krka estuary in Yugoslavia by Zutic and Legovic (1987) have shown that in a two-layer flow estuary, the chemical changes typical of the FSI can occur at the halocline between surface freshwater and the deeper saltwater. In the Krka estuary a film of organic matter at the FSI was found, composed of dissolved and insoluble organic matter, which

Figure 1.15 Typical estuarine distributions of conservative (linearly diluted) and non-conservative (not linearly diluted) chemical constituents and other properties in relation to salinity, with emphasis on the FSI (Freshwater/Seawater interchange). Salinity expressed as ‰NaCl, with linear scale > 1‰, and logarithmic scale < 1‰. Arbitrary scales are used for the concentration of sodium, chloride and calcium as examples of conservative ions. Examples of non-conservative distributions are shown for pH, speciation for dissolved cadmium (as %), organic flocculation, and total species diversity (arbitrary scales). (After Mantoura, 1987.)

contributed to the stability of the FSI and also provided a food source for organisms. It should be emphasised that the Krka estuary enters the Mediterranean, which is non-tidal, and such a stable film may not be found in the more variable conditions of tidal estuaries. The chemical changes which occur at the FSI contribute to the food supply of zooplankton, as is discussed in Chapter 2, and may also be responsible for changes in the composition of the estuarine fauna at low salinities, as discussed in section 1.5.

The waters flowing into estuaries will convey large quantities of oxygen into the estuary, and additional oxygen will be supplied through the surface of the water and by plant photosynthesis. However the many organisms living within estuaries, especially in the bottom deposits, rapidly consume the oxygen and thus many sediments are anoxic, except for a thin surface layer. Where excessive organic enrichment occurs the multiplicity of micro-organisms so produced may also consume all the oxygen within the water body as well.

The temperature regimes within estuaries are often more variable than adjacent waters due to the shallowness of estuaries which expose the waters to both heating and cooling, and also due to varied inputs of water to the estuary which may be at different temperatures.

A more detailed account of estuarine chemistry and the techniques used for analysing estuarine water is given by Head (1985), and here we shall restrict ourselves to the nutrient chemicals which may control the production of plant material, and thus serve as the foundation for the estuarine ecosystem.

Estuaries are generally rich in the nutrients needed for plant growth, especially nitrogen and phosphates, as the supply of these nutrients is continually being replenished by supplies from rivers, the sea and the land. Within the Ythan estuary, Scotland, for example, it has been found that the freshwater supplies about 70% of the nitrate and 80% of the silicate which flows into the estuary, whilst marine water supplies about 70% of the phosphate. The main source of particulate carbon in this estuary was the sea which supplied 10 × more than the rivers, and up to 95% of this material was in the form of detritus rather than as living plant material. The relative concentrations of nutrients and organic matter in the estuary compared to the adjacent river and sea is presented in Table 1.4.

The presence of large quantities of suspended sediment in estuaries may cause substantial modifications to the supply of nutrients for plant production. In the Tamar estuary, England, the phosphate supply is the main limitation on the growth of phytoplankton, but the suspended sediments in the water exercise some form of buffering action such that they

Table 1.4 Mean annual concentrations of nutrients and organic carbon in the Ythan estuary, River Ythan and adjacent North Sea. (After Leach, 1971.)

	River Ythan	*Ythan estuary*	*North sea (NE Scotland)*
Inorganic phosphate	1.1	0.2–2.0	0.2–0.6
Nitrate nitrogen	133.0	10–200	0.5–8
Reactive silica (all as μg-at. l^{-1})	206.5	10–220	1–5
Particulate organic carbon (mg l^{-1})	1.22	0.9–2.7	—
Particulate organic carbon (kg $tide^{-1}$)	345	—	1680–4600

remove phosphate from the water when the concentration is high and release it when it is low. Thus as the phytoplankton assimilate phosphate from the water, so there will be a tendency for more phosphate to be released from the sediment. The turbidity of estuarine waters may also limit phytoplankton production by restricting the penetration of light.

The distribution of sediments and salinity within estuaries, along with the distribution of oxygen, temperature and organic debris are all interdependent. The pattern of distribution of these factors varies from estuary to estuary depending on topography and the volumes of water involved. Inherent in the nature of true estuaries around the world are certain ecological common denominators, or factor complexes. They are: (1) the presence of well-aerated, constantly moving, relatively shallow water, which is mostly free from excessive wave action or rapid currents; (2) a salinity gradient from near zero up to over 30‰, and accompanying chemical gradients; (3) a range of sedimentary particle sizes from colloids to sands and detritus, resulting from land weathering, water transport, and estuarine processes; and (4) complex molecular interactions in both water and sediments, taking place in an abundance of dissolved and particulate organic matter, micro-organisms and fine sedimentary particles.

For the biologist looking at the estuarine ecosystem certain characteristic zones emerge, each with typical sediments and salinity. In particular estuaries one zone may occupy a large proportion of the area, and the other zones may be compressed, but a clear sequence can always be seen:

1. *Head* Where fresh water enters the estuary, and river currents predominate. Tidal but very limited salt penetration. Freshwater–saltwater interface (FSI). Salinity < 5‰. Sediments becoming finer downstream.

2. *Upper reaches* Mixing of fresh and salt water. Minimal currents, especially at high tide, leading to turbidity maxima. Mud deposition. Salinity 5–18‰.

3. *Middle reaches* Currents due to tides. Principally mud deposits, but sandier where currents faster. Salinity 18–25‰.

4. *Lower reaches* Faster currents due to tides. Principally sand deposits, but muddier where currents weaken. Salinity 25–30‰.

5. *Mouth* Strong tidal currents. Clean sand or rocky shores. Salinity similar to adjacent sea (> 30‰).

The relationship of these zones to the distribution of animals and plants will be discussed in section 1.5.

1.5 Distribution of estuarine organisms

Estuarine plants and animals fall into several categories as follows:

1. *Oligohaline organisms* The majority of animals living in rivers and other fresh waters do not tolerate salinities greater than 0.5‰, but some, the oligohaline species persist at salinities of up to 5‰.

2. *True estuarine organisms* These are mostly animals with marine affinities which live in the central parts of estuaries. Most of them could live in the sea but are apparently absent from the sea due to competition from other animals. Most commonly at salinities of 5–18‰.

3. *Euryhaline marine organisms* These constitute the majority of organisms living in estuaries with their distribution ranging from the sea into the central parts of estuaries. Each species has its own minimum salinity that it can live in. Many disappear by 18‰, but a few survive at salinities down to 5‰.

4. *Stenohaline marine organisms* These are marine organisms which occur in the mouths of estuaries, at salinities down to 25‰.

5. *Migrants* These animals, mostly fish and crabs, spend only a part of their life in estuaries, with some such as the flounder (*Platichthys flesus*)

Table 1.5 Classification of approximate geographic divisions, salinity ranges, types and distribution of organisms in estuaries. (From Carriker, in Lauff, 1967.)

Divisions of estuary	*Venice system* *Salinity ranges ‰*	*Zones*	*Ecological classification. Types of organisms and approximate range of distribution in estuary, relative to divisions and salinities*
River	< 0.5	limnetic	limnetic
Head	0.5–5	oligohaline	oligohaline
Upper Reaches	5–18	mesohaline	mixohaline
Middle Reaches	18–25	polyhaline	true estuarine
Lower Reaches	25–30	polyhaline	
Mouth	30–40	euhaline	stenohaline marine; euryhaline marine; migrants

or the shrimp (*Pandalus montagui*) feeding in estuaries, whilst others such as the salmon (*Salmo salar*) or the eel (*Anguilla anguilla*) use estuaries as routes to and from rivers and the open sea.

The distributions of these categories of estuarine organisms are summarised in Table 1.5.

Many studies of the distribution and abundance of animals and plants in estuaries have shown that the number of species within estuaries is less than the number of species within either the sea or fresh waters. If we consider the abundance of the various categories of estuarine organisms a clear pattern emerges as shown, for example, in data from the Tay estuary (Figure 1.16).

In the major groups of species living within estuaries a clear generalised pattern of declining diversity can be seen as one enters the estuary from either end. The number of oligohaline species is high in true fresh waters, but declines within estuaries. If the number of species within fresh water is expressed as 100%, then by 5‰ the number of oligohaline species has fallen to below 5%. The numbers of stenohaline marine species similarly falls as one enters estuaries from the seaward end, so that of the stenohaline species present in full strength sea water, virtually none are present below 25‰. The numbers of species of euryhaline marine organisms also declines progressively from the sea into estuaries, although in this case some survive down to 5‰. To offset this pattern of declining diversity of species, the numbers of species of true estuarine animals in the sea or fresh waters is low or nil, but increases within estuaries with a maximum number of species in the 5 to 18‰ region. However the total number of true estuarine species, as well as the number of migratory species, is relatively low compared to the

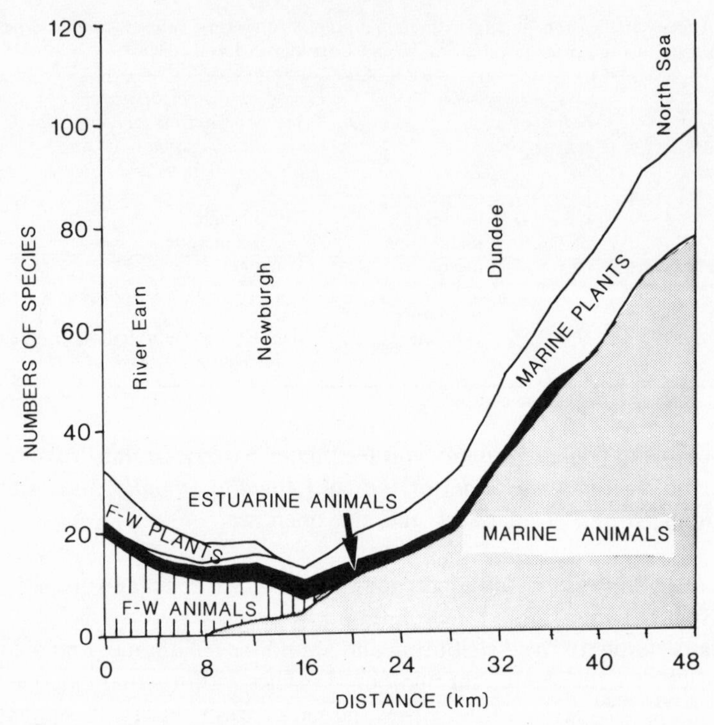

Figure 1.16 Composition of the fauna and flora along the Tay estuary, Scotland. Numbers of species, for each grouping, shown against distance. (After Alexander, Southgate and Bassindale, 1935.)

numbers of oligohaline or marine species. Thus when the numbers of species of all groups living within estuaries is combined in a single diagram, as first drawn by Remane, and known as 'Remane's curve' (Figure 1.17), it can be seen that estuaries are characterised by having fewer species than adjacent aquatic environments.

The fish fauna of estuaries for example may be classified into these categories:

	Number of fish species	
	Ponggol estuary Singapore	Forth estuary Scotland
Oligohaline	Absent	1
True estuarine	4	9
Euryhaline marine	9	7
Stenohaline marine	11	12
Migratory	2	5

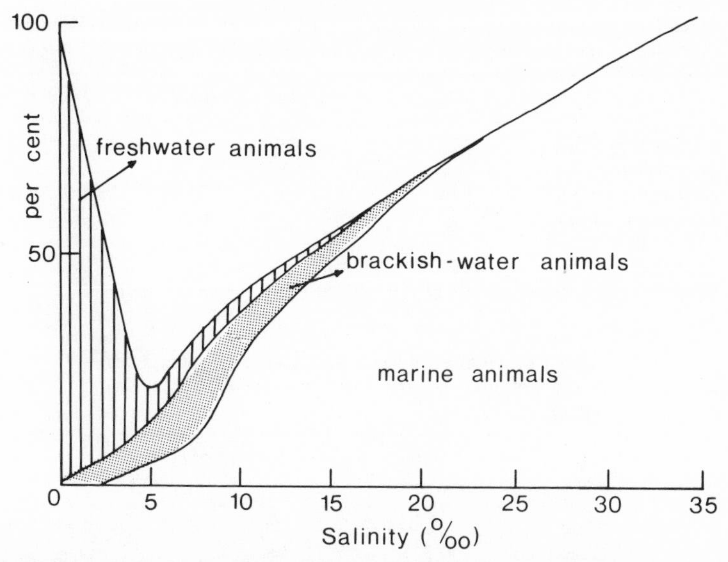

Figure 1.17 Generalised penetration of marine, freshwater and brackish-water animals into an estuary in relation to salinity. Diversity of marine and freshwater animals shown as a percentage of species diversity in each source habitat. Diversity of brackish-water animals shown as a sub-division of marine animals. (After Remane and Schlieper, 1958.)

In both of these cases the marine components can be seen as the dominant contributors to the diversity of estuarine species, here fish, and true estuarine and majority components contribute fewer species.

Data from a survey of the benthos of the Tay estuary (Scotland) shows a clear decrease in the number of species from the mouth of the estuary inwards (Table 1.6), controlled by two factors: salinity and substrate.

Table 1.6 Number of benthic species recorded from transects in the Tay estuary (after Khayralla and Jones, 1975).

Transect number	*Number of species*	*Substrate*	*Salinity (‰ NaCl)*
i (Buddon Ness)	51	sand	32–33
ii	53	sand	—
iii	36	sand	11–32
iv (Dundee)	23	mud/rock	6–30
v	22	sand	0.2–20.6
vi	21	mud/sand	—
vii	8	mud	0–10.7
viii	7	mud	—
ix (Newburgh)	7	mud/sand	0–0.3

Table 1.7 Faunistic components in the Avon-Heathcote estuary (New Zealand), expressed as species number (after Knox and Kilner, 1973).

	Mouth	*Middle*	*Head*	*Total*
Salinity range (‰)	25–34.3	8.4–30.3	0.2–15.7	
Freshwater	0	2	8	8
Estuarine	2	10	8	10
Euryhaline marine	69	67	10	72
Stenohaline marine	74	3	0	74
Total species	145	82	26	164

Table 1.8 Number of benthic macroinvertebrate species in the Potomac estuary. From data in Lippson *et al.* (1981).

Zone	*Salinity (‰ NaCl)*	*Number of benthic invertebrate species*
Tidal fresh	0–0.5	70
Oligohaline	0.5–5	104
Mesohaline	5–18	184
Polyhaline	18–30	199

Salinity determines the distance that a species is capable of penetrating into the estuary, but the full potential of any species to enter the estuary can only express itself when suitable substrates are present. Data from the Avon-Heathcote estuary (New Zealand) (Table 1.7), shows how small the estuarine and freshwater component of the fauna is, with 10 and 8 species respectively, compared to 146 marine species, of which 72 are stenohaline, and 74 are euryhaline. Data from the Potomac estuary is summarised in Table 1.8, where the sequence of species within the estuary can also clearly be seen.

In most estuaries there is a close connection between salinity distribution and substrate type, with reduced salinity associated with finer substrates. The close connection between these two physical factors, which has already been discussed, often makes it difficult for the biologist to distinguish their effect. Indeed a combined plot of the minimum salinity and the maximum silt content, as it affects individual estuarine species produces a unified curve (Figure 1.18). The ecological factors of salinity and substrate are thus closely interwoven in explaining the distribution of estuarine animals.

By contrast to the decline in the diversity of species within estuaries, it must be emphasised that the abundance of individual species often

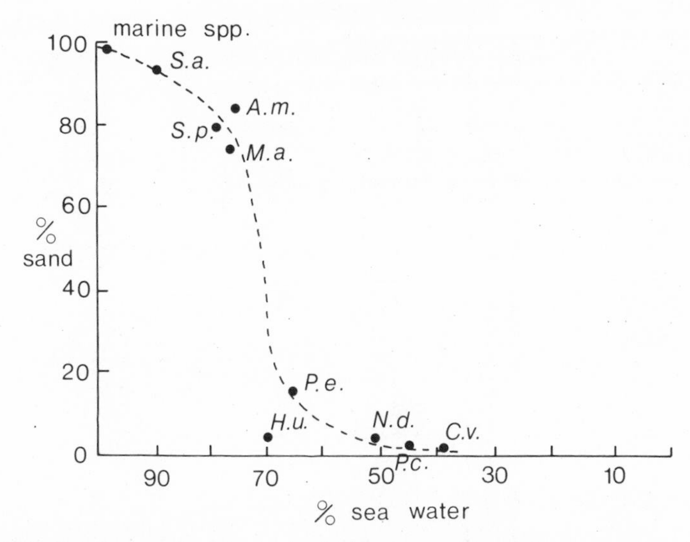

Figure 1.18 Distribution of species in the Ribble estuary, England, showing the minimum salinity of sea water and maximum silt content of substratum at which the species have been recorded. *S.a.* = *Scoloplos armiger; S.p.* = *Scrobicularia plana; A.m.* = *Arenicola marina; M.a.* = *Mya arenaria; P.e.* = *Pygospio elegans; H.u.* = *Hydrobia ulvae; N.d.* = *Nereis diversicolor; P.c.* = *Polydora ciliata; C.v.* = *Corophium volutator*. (After Popham, 1966.)

increases markedly within estuaries. Thus a picture emerges of a greater number of individual animals, but of fewer species. In Table 1.9 a comparison is made between the numbers and weight of animals living in the bottom deposits of marine beaches, fresh waters and estuaries at similar latitudes, from which it can be seen that the estuary has the highest number and weight of organisms.

In addition to the physico-chemical factors, discussed previously, which control the distribution and abundance of estuarine organisms, Wildish (1977) has identified three major biotic factors which exercise some control over the numbers of animals inhabiting estuaries: (1) food supply, (2) supply of colonising larvae, and (3) interspecific competition. The richness of the food supply, especially in the form of detritus, will be discussed in Chapter 2. The supply of larvae within estuaries does present some problems, as planktonic larvae may get carried out of the estuary by flushing currents. Many estuarine animals have thus suppressed the planktonic larval stage, and only have bottom-living larval forms. Competition occurs in all ecosystems, but I would suggest that the relatively harsh estuarine environment which challenges the physiological

Table 1.9 Comparison between numbers and weight of benthic animals from selected marine, estuarine and freshwater habitats in Britain. Data from McIntyre (1970), Buchanan and Warwick (1974), McLusky, Teare and Phizacklea (1980), Maitland and Hudspith (1974).

	Marine beach			*Marine sublittoral*	*Estuarine beach*		*Freshwater loch*
	exposed	*mod. exposed*	*sheltered*		*lower reaches*	*upper reaches*	
Number of species	8	12	19	40	18	4	54
Number of animals m^{-2}	450	600	1600	1000	18 000	100 000	58 000
Mean weight of animals g flesh dry wt m^{-2}	0.58	1.13	9.2	3.98	21.3	27.9	11.4

mechanisms of all species (see section 1.6) excludes many species, and thus the estuary represents for those species which can adapt to the environment, an escape from competition in the sea or fresh water. Thus those species which can adapt to estuaries are able to utilise'the rich food supply available there, and it is this food supply which becomes the main factor controlling the biomass and productivity of estuarine animals.

The estuarine habitat is thus not a simple overlapping of factors extended from the sea and from the land, but a unique set of its own physical, chemical and biological factors. The estuarine habitat has provided the environment for the evolution of some true estuarine organisms, but even more has provided a productive environment for those species which have entered it from the river, or more commonly from he sea. The estuarine habitat is thus characterised by relatively few species, but these species may be very abundant.

1.6 Problems of life in estuaries

Estuaries are characterised by having abundant populations of animals, but with relatively few species. What features within estuaries serve to restrict the diversity of species, and explain the observed decrease (Figures 1.16, 1.17) of species within estuaries?

Sanders (1969) developed a 'stability-time hypothesis' to compare the diversity of animals living within different bottom deposits of the sea and estuaries. In this approach the number of species present within a sample is plotted against the number of individual animals present in the same sample. When several environments are compared (Figure 1.19) a series of curves are produced. For estuarine samples the curve is always near the abscissa (lower axis) indicating fewer species in relation to the number of animals present than occurs with marine samples, which confirms the situation for estuaries of low diversity but high abundance. Sanders further suggests that estuarine environments are 'physically controlled communities', where physical conditions fluctuate widely and are not rigidly predictable. Thus organisms are exposed to severe physiological stress and/or the environment is of recent past history. By contrast, samples from the deep sea are described as 'biologically accommodated communities' with a large number of species per unit number of individuals, where physical conditions are constant and have been uniform for long periods of time. From such data the stability-time hypothesis (as shown in Figure 1.20) can be developed, with species number diminishing as physiological stress increases.

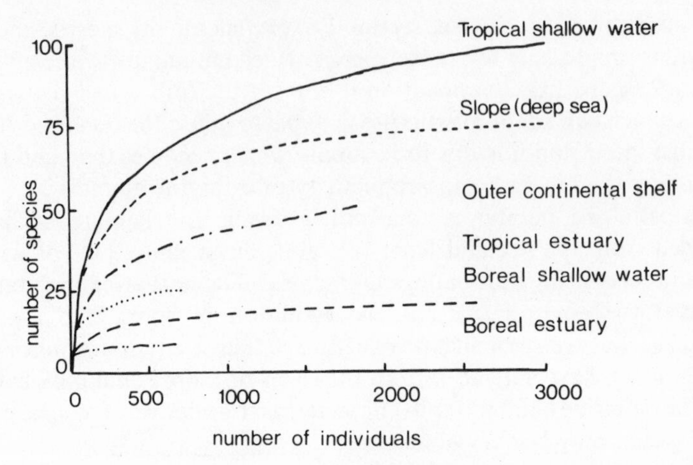

Figure 1.19 Arithmetical plot of the number of species present compared with the total number of individual animals present for sampling stations from different marine and estuarine habitats. (After Sanders, 1969.)

Two main hypotheses emerge to explain the paucity of estuarine species. The first explanation is that of physiological stress, and the second that the estuarine environment is of recent past history and that there has been insufficient time for a large number of species to develop within estuaries. These two hypotheses are not mutually exclusive and both could be operative.

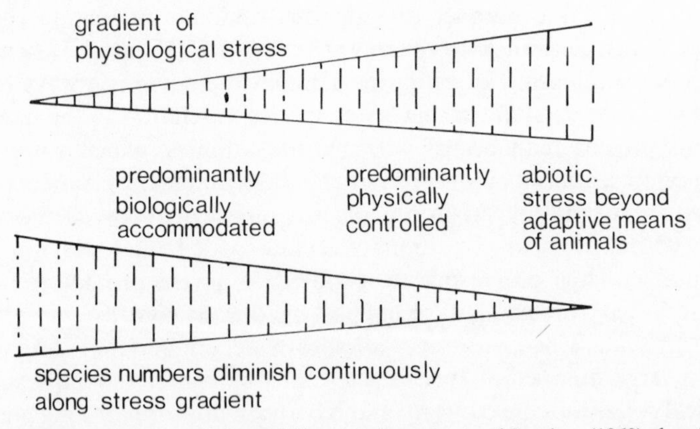

Figure 1.20 Representation of the stability-time hypothesis of Sanders (1969) showing that as stress increases so the number of species present decreases. As stress increases so the ecosystem comes under the control of predominantly physical factors until it finally becomes abiotic.

The general term 'physiological stress' covers a range of problems which confront the estuarine organism. The most conspicuous and probably best studied physiological factor is salinity, since estuarine organisms are exposed to variable and fluctuating salinities in contrast to the stable salinities which characterise both marine and freshwater habitats. Other physiological problems which impinge on the life of an estuarine organism are the nature of the substrate, both in terms of the fine particulate matter which can clog delicate organs, and of the virtually anoxic conditions which occur within muddy deposits. The responses to physiological stress can be many and varied, ranging from adaptations of behavioural patterns, and the modifications of particular organ systems, through to the evolution of new races or species adapted to the estuarine ecosystem. These physiologial adapations may take the form of the ability, or otherwise, of an animal to tolerate and thrive in the estuarine environment, such as the tolerance of particular temperature, salinity or substrate conditions. The most studied physiological and ecological factor is undoubtedly salinity.

The effects of salinity on estuarine organisms are varied. The impact of salinity, as with other environmental factors, is often multivariate. Thus for example temperature may interact with salinity, and the response of an animal to a change in salinity may be different at different temperatures. Salinity may affect an animal through changes in several chemical properties of the water. Firstly, the total osmotic concentration of sea water reduces as it is diluted with fresh water. Secondly, the relative proportion of solutes within estuarine water varies with the salinity, especially at very low salinities as described above. Thirdly, the concentration of dissolved gases varies with the salinity, with fresh water containing more oxygen than sea water at the same temperature. Finally, the density and viscosity of water varies according to salinity, with fresh water lighter than salt water. Thus when it is reported that an animal responds to a change in salinity it is important to check whether the animal is responding to the change in salt concentration, or to any of the above factors instead. The effects of salinity on an animal may also differ at different stages in the life cycle. As a general remark it appears that animals are more sensitive to extremes of salinity at the egg stage, when newly hatched, or when in adult breeding condition, than they are at intermediate stages of growth.

Salinity stress may invoke several responses. Confronted by an abnormal salinity an animal may seek to escape or otherwise reduce contact with the water. Such behavioural responses are common in many estuarine animals. For actively swimming animals, such as fish, it is relatively easy to escape from an adverse salinity, but sedentary animals such as barnacles can only respond by reducing contact and attempt to seal themselves inside their shell. For a burrowing animal it is often possible to retreat into the burrow, or even dig deeper, when confronted by an abnormal salinity, but as for the sedentary forms such a solution is only useful if the abnormal salinity is of a short duration.

If it is not possible to escape or reduce contact with the abnormal salinity, then the organism must rely on physiology responses. Either it can allow the internal environment (blood, cells, etc.) to become osmotically similar to the external environment (sea or estuarine water), or else it can attempt to maintain the internal environment at an osmotic concentration different from the external environment by the process of osmoregulation. For most stenohaline marine animals the former situation applies. When living in the sea their internal environment is isosmotic (osmotically similar) with sea water, thus although the ionic composition of the internal environment may be different from sea water, the total salt concentration is similar to the sea water. As such an animal enters an estuary the osmotic concentration of the internal environment falls to maintain equality with the less saline estuarine water. This passive tolerance of osmotic change is, however, only possible down to 10–12‰ because below this salinity the cells and tissues of the animal may cease to function.

As an alternative to the passive tolerance of osmotic dilution an animal may attempt to osmoregulate and maintain the internal environment at a different osmotic concentration from the outside. Thus for example an oligohaline, or true estuarine animal, which is living in salinities below 10‰, must keep the internal concentration at a level greater than the outside. Such an animal, termed a hyperosmotic regulator, will strive to maintain its blood concentration greater than 10–12‰, even though the external environment is lower, and thus enable the cells and tissues to function.

An animal which hyperosmoregulates when living in low salinities is faced with two options when living at high salinities. Either it can become isosmotic with the sea water, or it can seek to minimise the fluctuations of the external salinity regime by maintaining the blood less concentrated, or hypo-osmotic, than the external salinity. In the latter case the animal may thus live in salinities ranging from near zero to over 35‰, whilst the blood concentration may only rise from about 12‰ to 22‰. Such an animal, termed a hyper/hypo-osmoregulator can thus achieve a much more stable internal environment than is the case with a hyper/iso-osmoregulator which has still achieved greater stability than the completely iso-

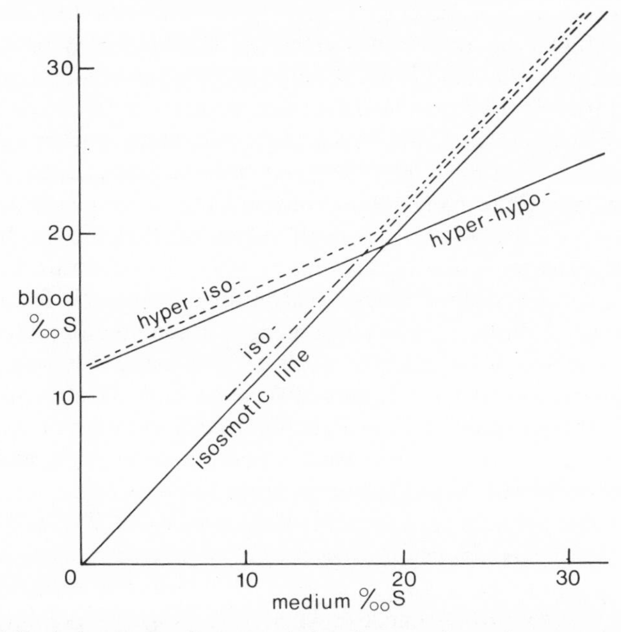

Figure 1.21 Typical patterns of osmoregulation. Iso-osmotic species are typically marine species. Typical hyper-iso-osmotic species are *Nereis, Corophium* or *Gammarus*, and typical hyper-hypo-osmotic species are *Crangon, Praunus* or *Neomysis*. The concentrations of the blood and the water medium are expressed as ‰ NaCl.

osmoregulator. These patterns of osmoregulation are presented in Figure 1.21. The mechanism of osmoregulation in animals relies on several possible physiological processes which are fully described by Rankin and Davenport (1981).

The most common feature of the physiology of estuarine animals is thus their ability to tolerate variations in their environment. With regard to salinity they are generally *euryhaline*, tolerating a range of salinities, in contrast to the *stenohaline* marine and fresh water animals which can tolerate only a small range of salinities. The ecophysiology of estuarine animals can in general be termed *euroyoecious*, that is living in a wide range of ecological conditions. However we should remember that only few of the many animals inhabiting the aquatic world have developed these abilities, and among those that have these abilities the pressures of competition cause different species to live in different parts of the estuary.

Estuaries offer excellent opportunities for the study of competition, coexistence and evolution of closely related species. The presence of closely related species in European estuaries is generally attributed to the brief geological history of present-day estuaries. Estuaries may be constantly changing due to the deposition of sediments, and to changes in sea level following the last Ice Age. Thus, the estuaries that we see today in Europe and America did not exist 10 000 years ago, and might be unrecognisable if we were able to visit them in 10 000 years time. Estuaries as environments have been present for many millions of years, but not those that most of us currently see and know. It has been suggested that the life history of individual estuaries is too short for many species to colonise and adapt to them, and this has been given as a reason for the relatively few species in them today. Further, the environmental gradients that characterise estuaries enables us to examine what factors cause one species to be ecologically and genetically separated from another. Deaton and Greenberg (1986) have suggested that euryhaline animals may have lower rates of speciation than marine or freshwater species, but the many examples of speciation presented below do not conform to this view.

Within the polychaetes there are three species of *Nereis* present in European estuaries, namely *N. diversicolor, N. virens* and *N. succinea. N. virens* is larger and more voracious than the others, but intolerant of low salinities. *N. diversicolor* can live in a wide range of salinities, but is apparently excluded from higher salinities by competition from *N. virens. N. succinea* forms an intermediate species, but may disappear completely with low temperatures.

The two main *Corophium* species in estuaries appear to be separated by both substrate and salinity. *C. volutator* is more tolerant of low salinities, whilst *C. arenarium* is more abundant at high salinities and with sandier substrates.

There are four similar Hybrodiid snails in estuaries. *Potamopyrgus jenkinsi* is found typically in fresh water, and in estuaries with salinities up to 15‰. *Hydrobia ventrosa* is found between 6 and 20‰, *H. neglecta* from 10 to 24‰, and *H. ulvae* from 10 to 35‰. However, salinity is not a complete explanation, as *H. ventrosa* and *H. neglecta* occur mainly in areas of stable salinities such as the Baltic Sea and the Belt Sea, whereas *H. ulvae* dominates in the fluctuating salinities that typify many estuaries.

Four species of *Nephtys* worms occur in the estuarine regions of the Netherlands, namely *N. caeca, N. cirrosa, N. hombergi* and *N. longosetosa.* These species inhabit at least three separate ecological niches. *N. caeca* occurs subtidally in muddy sands, whilst *N. hombergi* lives both intertidally and subtidally in even muddier substrates. *N. cirrosa* and *N. longosetosa* both inhabit clean sands. Thus in these species it can be seen that substrate preference acts as the main factor isolating one species from another.

Two species of cockle occur in European estuaries, *Cerastoderma edule* (= *Cardium edule*) and *Cerastoderma glaucum* (= *C. lamarcki*). In Danish estuaries *C. glaucum* is the principal cockle at salinities below 18‰, whilst both species occur at salinities of 20–25‰. At 30–35‰, *C. edule* is the principal species, but at salinities of 60‰ in the Sea of Azov *C. glaucum* occurs. It thus appears that the 'brackish-water cockle' *C. glaucum* can tolerate a range of salinities wider than *C. edule*, but in normal sea water the 'edible cockle' *C. edule* is the dominant species.

The importance of osmoregulation and salinity tolerance as a means of limiting the penetration of particular species into estuaries has been shown in a study of the isopods of *Jaera albifrons* group. *Jaera ischiosetosa* can survive for several days in fresh water and is able to maintain a steady internal osmotic pressure even in very dilute sea water, and is the species which penetrates furthest into estuaries. *Jaera forsmanni* and *Jaera praehirsuta* have lower survival rates in dilute salinities and are less capable of maintaining a steady state in dilute sea water, and are found at the seaward end of estuaries. *Jaera albifrons* is intermediate in both osmoregulatory ability and in its ecological distribution. These physiological differences, coupled with the observed ecological differences show how genetic isolation between the species can occur.

Despite these examples of active speciation amongst estuarine animals, it should be emphasised that the distribution of fauna across an estuarine mudflat may be quite stable. Mettam (1983) showed that the infauna on an intertidal area in the Severn estuary showed little change over a period of 40 years, and in the Ruel estuary (Loch Riddon) McLusky and Hunter (1985) also showed little change over a period of 53 years, which in both cases was attributed to the constancy of the sediments of the area.

In considering the possible causes for the brackish water minimum of species at 5‰, the so-called Remane's minimum shown in Figure 1.17, Wolff (1973) and Deaton and Greenberg (1986) reject pollution, size of estuaries, turbidity or the critical salinity hypothesis of Khlebovich, as satisfactory explanations. Pollution is rejected as an explanation because the decrease in species within temperate estuaries occurs in both unpolluted and polluted estuaries alike, although the decrease will be more marked in polluted situations (Chapter 5). The size of estuaries as a reason is rejected because the decrease in species occurs in both small coastal estuaries as well as large brackish seas, such as the Baltic. Furthermore, freshwater bodies have greater species diversity than larger estuaries nearby. The turbidity of estuarine waters may inhibit marine species with delicate gill structures, but the decline of species in brackish waters occurs in both turbid estuaries, as well as estuaries with lower sediment loads, and non-turbid brackish seas. The critical salinity hypothesis of Khlebovich (1968) suggested that for many purposes estuarine and brackish waters with more than 5‰ salinity may be regarded as dilutions of sea water maintaining a constant ionic ratio, but that below 5‰ many chemical properties of estuarine water change. Accordingly Khlebovich suggests that these chemical peculiarities of estuarine water are responsible for the observed eco-physiological barrier at 5‰ dividing the marine and freshwater components of the estuarine fauna. This hypothesis is however invalidated by many important estuarine animals which live across the 5‰ divide, by the fact that the reduction in species occurs progressively all the way from the mouth of an estuary, and by the fact that the major ionic perturbations occur at very low salinities, the FSI at < 1‰, at which salinity species diversity is increased compared to 5‰.

If the above reasons are rejected as unsatisfactory explanations for the reduction of species in brackish and estuarine waters, what can be suggested? The main hypothesis which remains is that of physiological stress. The salinity conditions of estuaries present, as we have already seen, many challenges to the physiological processes of the estuarine habitants. The main challenge of living in a range of salinities rather than the constant salinities of the sea or fresh waters is met by the range of adaptations both genetic and behavioural that occur in brackish-water species. It appears that few of the many animals living in the sea have these osmoregulatory abilities, and the decline of species within estuaries can be seen as a progression of species each reaching their own physiological limit. At the seaward end are the many species which are iso-osmoregulators. However these each reach a minimal salinity determined by their own cellular tolerances. Thereafter, estuaries are inhabited by hyper- and hypo-osmoregulators, but the extent of their osmoregulatory powers varies from species to species, and so a decrease in species occurs. The rate of decrease of species is generally more marked in estuaries than in brackish seas. This difference in the rate of decrease may be explained by the stability of the salinities in each area. In a brackish sea, such as the Baltic, there is a progressive decrease in salinity, and an accompanying decrease in species diversity, but many species are able to live at lower salinities in the stable low salinities of the Baltic than they are able to in the salinity regimes of estuaries which vary with each tide as well as seasonally. It thus appears that the fluctuations of salinity within estuaries enhances the rate of decrease of species, but since a decrease occurs in both stable as well as fluctuating situations instability cannot be an exclusive explanation.

1.7 The estuarine food web

The estuarine food web is dependent on the input of energy from sunlight and the transportation of organic matter into the estuary by rivers and tides. Within the estuary the plants, or primary producers, convert these inputs into living biological material. As the plants grow they are consumed by the herbivores, or primary consumers, which are in turn consumed by the carnivores, or secondary consumers. In order to understand the conversion of material as it passes through the links of the estuarine food web we need to study and quantify the life of each member of the food web.

The simplest means of quantification is to identify, count and weigh the organisms within particular parts of the estuary. The identification of the organisms is of course the first step, and can generally be performed with the aid of guides to the sea-shore which can be found in most countries. The estimation of the number of organisms depends on choosing a unit area, usually 1 m, and counting the animals and plants within the area. For small organisms it is usually impracticable to count everything in 1 m^2, so one would count those within a smaller area such as 10 cm^2 and multiply the result by an appropriate factor to produce a result in terms of 1 m^2. Such counts need to be repeated several times and the mean result determined. If the animals are buried in the sediment, as will be the case in many estuarine species, then the sediment must be sieved through an appropriate sieve to extract the animals. A 1 mm sieve is suitable for bivalve molluscs, but for most other estuarine animals a 0.25 mm sieve is needed to be certain that a true estimation of abundance is obtained. Baker and Wolff (1987) give fuller details of field methods for sampling estuarine organisms.

The typical results of an exercise in counting estuarine animals is shown in Figure 1.22, from which a clear pyramid of numbers can be seen, with many small animals, decreasing to fewer larger animals. If however the animals are weighed the reverse picture can be seen, Figure 1.23, with the total weight of the smallest animals considerably less than the total weight of the larger animals, the pyramid of weight. The name given to the weight of the organisms is the *biomass*. The measurement of biomass alone, important though it may be in comparing the immediately available standing crop from one part of an estuary to another part, is quite inadequate for the purpose of estimating the amount of food which one

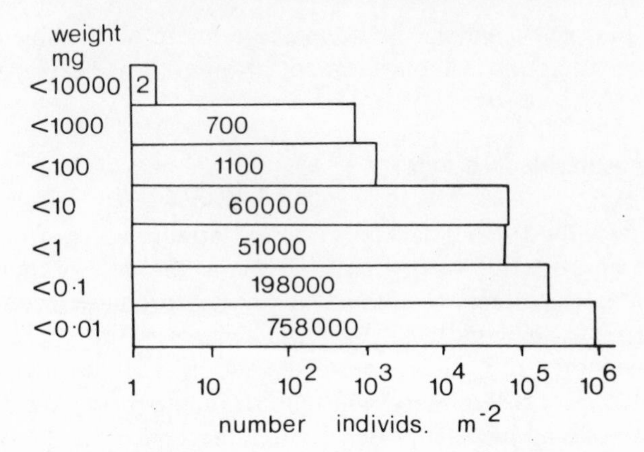

Figure 1.22 The pyramid of numbers. The number m^{-2} of the fauna of Niva Bay, Denmark, divided into seven weight categories. (After Muus, 1967.)

trophic level, such as the plants, can make available to the animals feeding on them. In order to understand these feeding, or trophic, relationships we need to know the rate of production of organic matter by the various members of the estuarine ecosystem. The distinction between the rate of production (or productivity) of organic matter by an organism or community, and the biomass (or standing crop) consisting of the organism itself or the community, is fundamental.

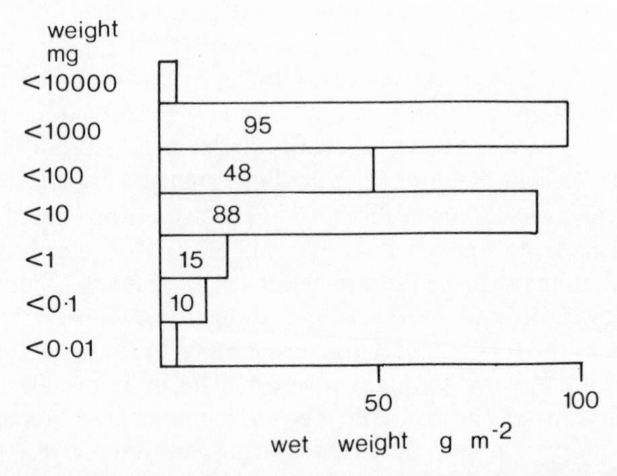

Figure 1.23 The pyramid of weight. The weight m^{-2} of the fauna of Niva Bay, Denmark, divided into seven weight categories. Compare with Figure 1.22. (After Muus, 1967.)

The most comprehensive form that investigations of the rate of production can take are the measurement of the flow of organic matter and energy through each component of the ecosystem. The principles of energy flow in ecosystems are well described by Phillipson (1966), and their application to aquatic benthic organisms are especially considered by Crisp (1971). A summary of the terms and units used in productivity studies is given in Table 1.10.

For practical purposes the two most important terms when considering energy flow in ecosystems are the biomass and the annual productivity. Biomass gives information on the amount of biological life living in any area, and should always be expressed in terms of the living components only, omitting non-living parts such as calcareous shells, or in terms of

Table 1.10 Terms used in productivity studies which are utilised in this book (after Crisp, 1971).

Term	*Abbreviation*	*Definition*
Biomass	B	Amount of living substance constituting the organism under study
Mean biomass	$\bar{B}$	Mean amount of living substance constituting the organism during the time of study (typically one year)
Consumption	C	Total intake of food or energy
Egesta	F	That part of the consumption that is not absorbed but is voided as faeces
Absorption	Ab	That part of the consumed energy that is not voided as faeces
Excreta	U	Absorbed consumption that is passed out of the body as secreted material, e.g. urine, mucus
Assimilation	A	That part of the consumption that is used for physiological purposes (C excluding U and F)
Production	P	That part of the assimilated energy that is retained and incorporated in the biomass of the organism. May be regarded as 'growth'
Respiration	R	That part of the assimilated energy that is converted into heat directly or through work
Gonad output	G	Energy released as reproductive bodies
Yield	Y	That fraction of production utilised by another organism

Assuming the conservation of energy we may derive the following equations:

Consumption $C = P + R + G + U + F$ = (total energy budget)
Absorption $Ab = C - F = P + R + G + U$
Assimilation $A = C - (U + F) = P + R + G$

Derived from the above is the $P/\bar{B}$ ratio, which is the ratio of the production of an organism to the mean biomass of that organism over one year

Table 1.11 Units used in productivity studies which are utilised in this book (see also Crisp, 1971).

Term	*Abbreviation*	*Explanation*
Abundance	Number m^{-2}	The number of organisms contained within 1 m^2 either on the surface of the m^2 or in the substrate or water column underlying one m^2
Biomass (1)	g total wt	The total weight in grams of the living wet organisms
Biomass (2)	g dry wt	The weight of organisms after drying at 60–100°C (approx. 17–7% of (1))
Biomass (3)	g flesh dry wt	The dry weight of living animal which remains after a non-living shell has been removed (applicable to hermit crabs, bivalve molluscs, etc.)
Biomass (4)	g ash-free dry wt	The weight of an organism which has had a non-living shell removed, the water evaporated off and the ash weight of inorganic matter determined by ignition of the sample at about 450–600°C. The result is regarded as the best measure of the weight of living material
		Biomass may be expressed as energy units:
Calories	cal	*Definition:* Quantity of heat required to raise the temperature of 1 g of water through one centrigrade degree at 15°C
		Application: The calorific content of a biological sample can be determined in a bomb calorimeter and represents the energy released during complete combustion of 1 g of sample
Joule	J	SI unit which has replaced the calorie
		1 cal = 4.184 J
		The results of energy units are usually multiplied 1000 fold and expressed as kcal or kJ g^{-1} (or kcal/g)
Gram carbon	gC	The carbon content of a sample can be determined and the biomass expressed in terms of carbon weight
Gram nitrogen	gN	The nitrogen content of a sample can be determined and the biomass expressed in terms of nitrogen weight

Biomasses expressed as gC or gN, like g ash-free dry wt, are intended as measures of the living tissue present in a sample, and provide a useful alternative to the determination of the energy content of the material.

The above biomasses (1–4) can also be expressed in kg or mg. However the values derived take no account of the energetic value of the tissue which may vary from species to species.

energy units. The derivation of various units of biomass and its conversion into energy units is given in Table 1.11. A summary of approximate conversions from one unit to another is given in Table 1.12.

The production of a population may be derived by one of two approaches. Either all the growth increments of all the members of the population under study are added for one year, or else growth is ignored and production is derived from the addition of the change in the biomass

Table 1.12 Conversion of units to and from each other. Conversions of units of biomass into units of energy vary from species to species, and from season to season, however certain approximations can be used as indicated below. (Data from Winberg, 1971; Crisp, 1971; Chambers and Milne, 1979; Salonen *et al.*, 1976.)

1 gC approx. = 10 kcal	
1 gC approx. = 2 g ash-free dry wt	
1 g ash-free wt approx. = 21 kJ	
1 g organic C approx. = 42 kJ	
1 litre oxygen = 4.825 kcal (oxycalorific equivalent)	
1 g carbohydrate approx. = 4.1 kcal	
1 g protein approx. = 5.65 kcal	
1 g fat approx. = 9.45 kcal	
1 cal = 4.184 J	
Some selected examples:	
1 g flesh dry wt *Cardium edule*	= 18.57–19.46 kJ
1 g ash-free dry wt *Cardium edule*	= 19.99–21.50 kJ
1 g flesh dry wt *Corophium volutator*	= 13.26–16.82 kJ
1 g ash-free wt *Corophium volutator*	= 18.66–20.88 kJ
1 g flesh dry wt *Mytilus edulis*	= 19.95–20.83 kJ
1 g ash-free dry wt *Mytilus edulis*	= 21.71–22.51 kJ
1 g flesh dry wt *Macoma balthica*	= 16.10–19.04 kJ
1 g ash-free dry wt *Macoma balthica*	= 19.38–21.46 kJ
1 g flesh dry wt *Nereis diversicolor*	= 16.44–19.70 kJ
1 g ash-free dry wt *Nereis diversicolor*	= 20.87–22.42 kJ
thus 1 g flesh dry wt estuarine invertebrate approx. = 18 kJ	
thus 1 g ash-free dry wt estuarine invertebrate approx. = 21 kJ	

over one year plus the mortality due to predation etc. over the same period. A clear description of the application of these alternative approaches is given by Crisp (1971). An alternative to these approaches has been given by McNeill and Lawton (1970) who showed that annual production can be calculated from a knowledge of the annual respiration of the population.

An energy budget for an intertidal population of the bivalve *Scrobicularia plana* may be taken as an example of the utilisation of energy by an estuarine invertebrate (Table 1.13). Note that 79% of the energy assimilated was utilised in respiration, which is a typical value for energetics studies, and only 2% of the energy consumed by *Scrobicularia* was consumed by the next trophic level.

The relationship between production and mean annual biomass is known as the turnover ratio, or $P/\bar{B}$ ratio. Waters (1969) compared various temperate benthic invertebrates and showed that the $P/\bar{B}$ ratio varied only between 2.5 and 5, with a mode of about 2.5. This value shows that production over a study year is 3 times greater than the mean biomass, and

Table 1.13 Energy budget for intertidal *Scrobicularia plana* (Data from Hughes, 1970).

Mean annual biomass	$\bar{B}$	203 kcal m^{-2}	
Change in biomass	ΔB	40 kcal m^{-2} yr^{-1}	
Mortality	E	20 kcal m^{-2} yr^{-1}	
Production	P	124 kcal m^{-2} yr^{-1}	
Assimilation	A	600 kcal m^{-2} yr^{-1}	
Gamete Production	G	64 kcal m^{-2} yr^{-1}	
Respiration	R	476 kcal m^{-2} yr^{-1}	
Egesta	F	388 kcal m^{-2} yr^{-1}	
Consumption (A + F)	C	988 kcal m^{-2} yr^{-1}	
$P/\bar{B}$ ratio		0.61	
Assimilation efficiency		A × 100/C	61%
Net growth efficiency		P × 100/A	21%
Ecological efficiency		E × 100/C	2%

emphasises that production is a much more important measure of the transfer of energy within an ecosystem than is the biomass. As we shall see in later chapters the $P/\bar{B}$ of estuarine animals from temperate areas may vary from less than 1, to well over 5. The significance of this ratio is that it is a clear indication of the ecological performance of a population. Two populations may for example have similar biomasses, but different $P/\bar{B}$ ratios, thus the one with the high ratio will produce more organic material to be made available to predators than the one with the lower ratio. In comparing different animals at one site, smaller animals will have higher $P/\bar{B}$ ratios than large animals, and Schwinghamer *et al.* (1986) have shown that reasonable estimates of annual production and its distribution among size groups in benthic communities can be derived from a knowledge of biomass and the size structure of the community (Figure 1.24).

A summary of $P/\bar{B}$ ratio values in the literature has been given by Ansell *et al.* (1978) for bivalve molluscs and other invertebrates inhabiting soft substrates. The ratio $P/\bar{B}$ as a basis for comparison suffers from the disadvantage that it gives high values for young animals of a species and increasingly smaller values for older animals. Thus comparisons of this ratio are only really valid where the population on which they are based have the same or similar age structure relative to their life span. The $E/\bar{B}$ ratio (elimination or mortality to biomass) or the mortality coefficient are therefore used as alternatives. The mortality coefficient, z, is roughly equivalent to the $E/\bar{B}$ ratio with E calculated on a daily basis. Nevertheless much published work on estuarine ecosystems remains published as $P/\bar{B}$ values, and results of such studies will be described in this book in their original units. A comparison of the range of values of production or

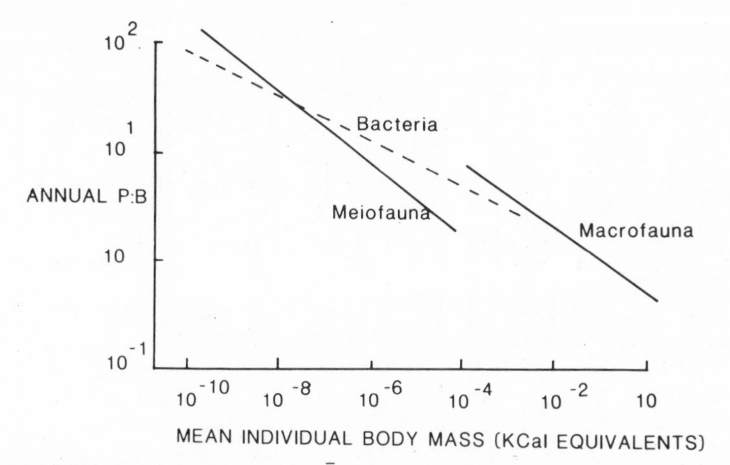

Figure 1.24 Relations between annual P: $\bar{B}$ (production: biomass ratio) and mean individual body mass (as kilocalorie equivalents) for benthic organisms. Geometrical mean functional regressions are shown for bacteria, meiofauna and macrofauna. (After Schwinghamer *et al.*, 1986.)

elimination: mean biomass ratios is given in Figure 1.25 where it can be seen that values of 0.10 to 5 are typical for animals from the Mediterranean and Caribbean, and 5 to 100 have been found for Indian species.

Organisms within the estuarine ecosystem may be assigned to trophic (or feeding) levels following the trophic–dynamic approach to ecology expounded by Lindemann (1942). The ecosystem may be defined formally as the system composed of physical–chemical–biological processes active within a space–time unit of any magnitude, and thus includes the biotic community of organisms plus its abiotic environment. Within the biotic community we can assign organisms to trophic levels, which are the groups of organisms which share a common method of obtaining their energy supply. The first trophic level are the producers, which are the plants obtaining their energy by photosynthesis. Following them are the primary consumers, which are herbivorous animals feeding on plant material. Following these are the secondary consumers, which are carnivorous animals feeding on the primary consumers. Eventually we reach the tertiary consumers feeding on the secondary consumers. At each stage in this trophic sequence energy is consumed, and whilst some of it is rejected as waste, or converted into bodily growth, the majority of the energy is dissipated as heat following respiration. The loss of energy due to respiration is progressively greater for higher levels in the trophic sequence, as is the efficiency of utilisation of the food supply. Due to the losses of

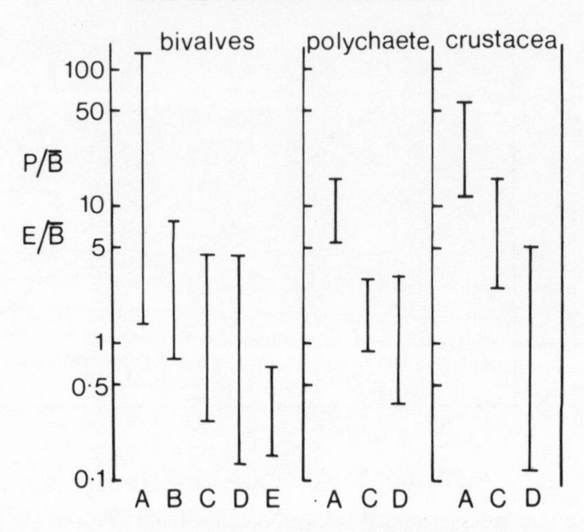

Figure 1.25 The range of values of production: mean biomass ratio (or elimination: mean biomass ratio) found in the literature for benthic invertebrates from different geographical areas. A = Indian tropical, B = Mediterranean, C = Caribbean, D = Northern temperate and Boreal, E = Arctic. (After Ansell *et al.*, 1978.)

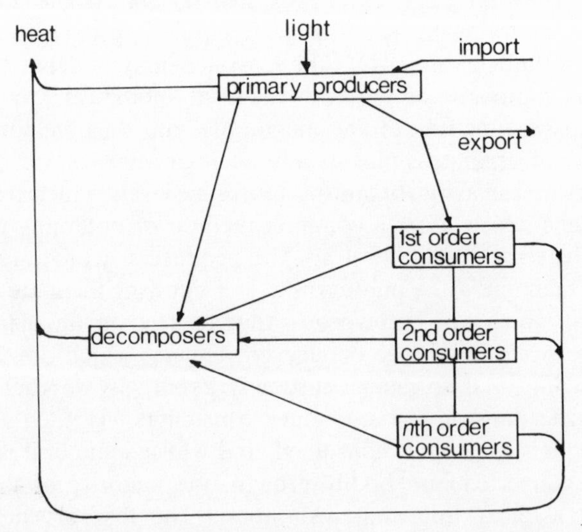

Figure 1.26 Model of trophic transfer in an ecosystem, as first modelled by Lindemann (1942).

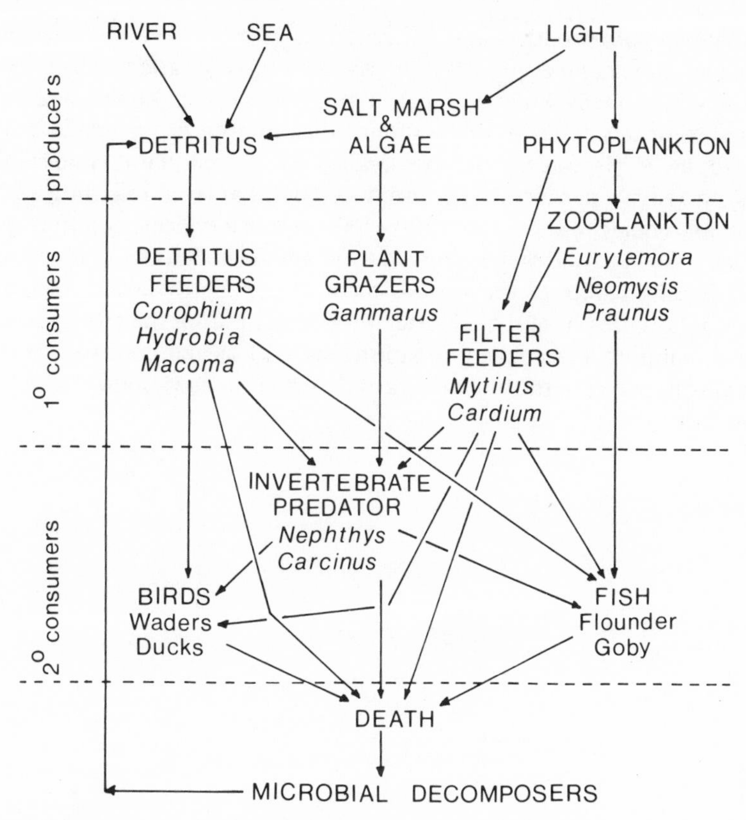

Figure 1.27 An estuarine food web. The arrows indicate the feeding relationships of a typical northern hemisphere estuary. The dotted lines indicate the division into producers, primary consumers and secondary consumers used in this book.

energy in food chains there are rarely more than five trophic levels. Apart from the loss of energy from each trophic level due to respiration and predation, other materials such as excreta or dead organisms pass to the decomposer trophic level where micro-organisms break down the material until it too is lost as heat or else reutilised as detritus by the detritivores which are regarded as members of the primary consumer level.

The relationship between these trophic levels in a simplified ecosystem is shown diagrammatically in Figure 1.26. It should be noted that the Lindemann model of trophic transfer is essentially a food chain with one level linking directly to the next. In practice when examining ecosystems we are looking at a food web with certain consumer animals for example feeding on both producer and different consumer levels of prey. Nevertheless the Lindemann approach is an essential basis for attempting to make sense of the many links within a natural ecosystem. In the succeeding chapters of this book we shall examine the estuarine ecosystem as a series of trophic levels, but the reader should be cautioned that it is not always easy to assign a particular organism to a particular level. In Figure 1.27 an attempt is made to show some of the links within a typical north European estuary. Such a food web is derived from an understanding of the biology and feeding habits of the various members of the estuarine ecosystem. Throughout this book I shall attempt to explain both the energy sources which support a particular organism, as well as the contribution that organisms makes to the functioning of the estuarine ecosystem of which it is a member.

CHAPTER TWO
PRIMARY PRODUCERS
PLANT PRODUCTION AND ITS AVAILABILITY

2.1 Introduction

Within the estuarine ecosystem there may be several sources of plant production. Growing on the intertidal zones are usually a number of salt marsh plants. In most European estuaries the salt marsh plants are confined to the topmost part of the intertidal zone where they are not covered by the tide every day, but in many American estuaries the salt marsh plants may occupy the major part of the intertidal area and be immersed at each tide. In other parts of intertidal zone may often be found the eel-grass (*Zostera*), which is a true flowering plant, or representatives of the algae. Some of the algae are attached to rocky outcrops such as the typical seaweeds, e.g. *Fucus* species. Also growing directly on the surface of the mudflats may be the filamentous algae, *Enteromorpha* species, or the single-celled microphytobenthos (also known as epibenthic algae). Within the water body are found floating members of the phytoplankton, for example diatoms or dinoflagellates.

The production of all these various plants is of course dependent on both sunlight and temperature, and may also be potentially limited by the availability of nutrients, especially nitrogen and phosphorus. These nutrients are typically rich in estuarine waters, having been carried there from the sea, rivers or land adjacent to the estuary. Within the estuary the nutrients are utilised by the plants, and following the death of the plant become recycled (Figures 2.1, 2.2) by the processes of decomposition to be utilised again by the plants. High levels of primary production occur in estuaries in comparison to the sea, due mainly to the high nutrient levels pertaining there (Table 2.1).

In considering the role of the primary producers as food sources for the primary consumers of the estuarine ecosystem it is necessary to consider the importance of detritus. Detritus has been defined by Darnell (1967) as ‘all

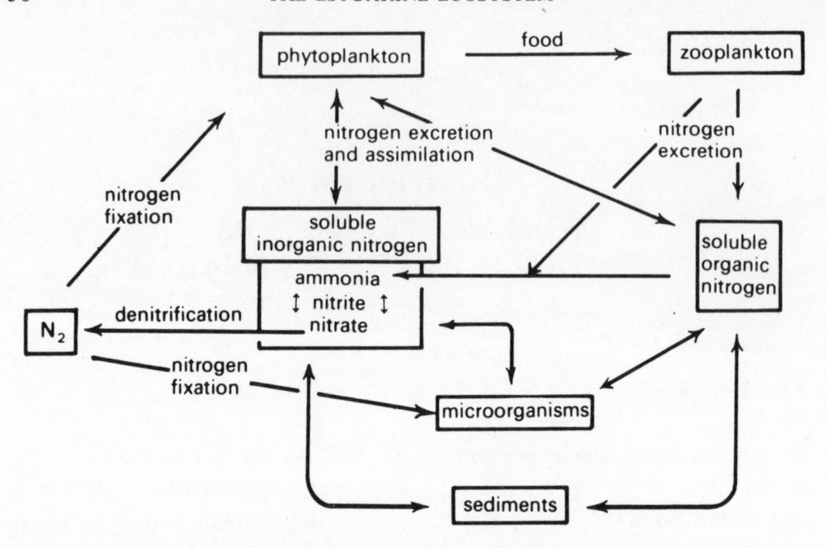

Figure 2.1 The nitrogen cycle in the sea. The exchange of inorganic and organic nitrogen with sediments is not fully understood. (From Meadows and Campbell, 1987.)

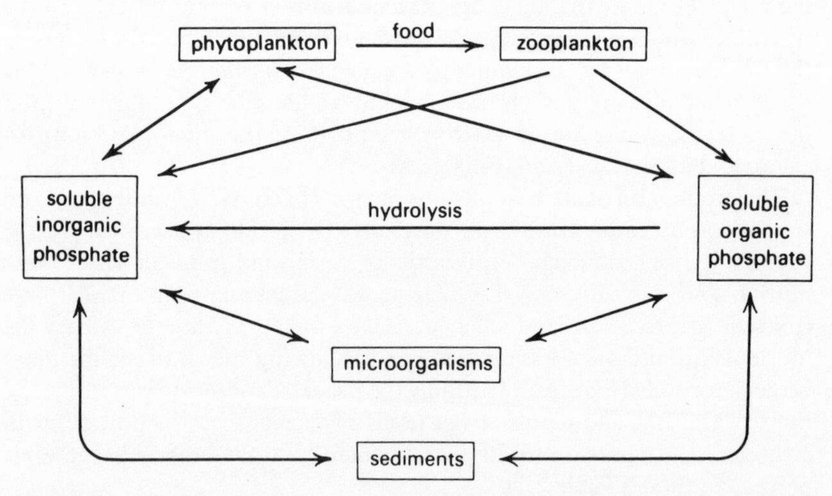

Figure 2.2 The phosphorus cycle in the sea. Zooplankton excrete inorganic and organic phosphate, and phytoplankton both excrete and assimilate inorganic and organic phosphate. The exchange of phosphate with sediments is not fully understood. (From Meadows and Campbell, 1987.)

Table 2.1 Net primary production in various marine habitats. Data from Strachal and Ganning (1977).

	Open sea	*Coastal zone*	*Upwelling regions*	*Estuaries (and coral reef)*
% sea	90	9.4	0.1	0.5
Net primary prod. $gC\ m^{-2}\ yr^{-1}$	50	100	300	1000
Net primary prod. 10^9 ton C yr^{-1}	16.3	3.6	0.1	2.0

types of biogenic material in various stages of microbial decomposition, which represents a potential energy source for consumer species'. Much of this biogenic material is fragments of plant material. Strictly speaking the bacteria and other microbial organisms which live on and decompose the plant fragments are a second trophic level, dependent on the first trophic level, the plants. However the phytoplankton, benthic microalgae, plant fragments and their decomposers become so intertwined, that the food for the primary consumer animals is generally called 'particulate organic matter', without regard to its exact origin.

In this chapter therefore we shall examine both the primary production of the salt marshes and algae (benthic or planktonic), and the limitations placed upon this productivity by nutrient availability, and also the fate of the plant material as it is fragmented and decomposed, and thereby becomes more available to consumer animals.

2.2 Salt marshes

The salt marsh as an ecosystem is well described by Long and Mason (1983), so we shall confine ourselves here to the contributions that salt marshes make to the total estuarine ecosystem.

The estuaries of the south-eastern coast of America are dominated by large stands of the marsh grass, *Spartina*, especially *Spartina alterniflora*, which may occupy up to 90% of the intertidal area. These salt marshes have long been recognised as being among the most productive ecosystems in the world (Table 2.2). Maximum production (up to 3300 g dry wt $m^{-2}\ yr^{-1}$ of above-ground material) occurs in southern US states, and this decreases northwards. The high overall levels off production are attributed to the ample supply of dissolved nutrients, coupled with a long growing season

Table 2.2 Net primary production of selected estuarine habitats, expressed as g C or g dry wt m^{-2} yr^{-1} (Data from various sources, see refs.)

	Locality	*g C*	*g dry wt*	*Source*
PHYTOPLANKTON	Lynher, UK	81.7		Joint (1978)
	Beaufort, USA	52.5		Williams (1972)
	Baltic Sea	48.94		Eriksson *et al.* (1977)
	Grevelingen NL	130		Wolff (1977)
	Barataria Bay, USA	210		Wolff (1977)
	Great South Bay, NY, USA	450		Boynton *et al.* (1982)
	Average for 45 estuaries	190		Knox (1986)
MICROBENTHIC ALGAE (see also Table 2.4)	Lynher, UK	143		Joint (1978)
	Grevelingen, NL	25–37		Wolff (1977)
	Barataria Bay, USA	240		Wolff (1977)
	Massachusetts, USA	105		van Raalte *et al.* (1976)
	Average for 15 estuaries	100		Knox (1986)
ZOSTERA	Leaves, Denmark		856	Sand-Jensen (1975)
	Below ground, Denmark		241	Sand-Jensen (1975)
	Range	58–330	116–680	Mann (1982)
THALASSIA		825		Greenway (1976)
SALT MARSH	*Limonium*, UK		1050	Jefferies (1972)
	Salicornia, UK		867	Jefferies (1972)
	Baltic meadow		230	Jefferies (1972)
	Oregon, USA		1200–1700	Eilers (1979)
	Carex, Oregon, USA		2600	Eilers (1979)
	Netherlands	100–500	200–1000	Wolff (1977)
SALT MARSH (*SPARTINA*)	Georgia, USA		3700	Jefferies (1972)
	N. Carolina, USA		650	Jefferies (1972)
	UK		970	Jefferies (1972)
	Range	133–1153	500–3500	Knox (1986)
MANGROVE	Range	0–2700		Knox (1986)
COMPARISONS with the SEA	*Laminaria*	1200–1800		Mann (1982)
	Coastal plankton	100		Mann (1982)
	Open sea plankton	50		Mann (1982)

and hybrid vigour displayed by the *Spartina* plants. However, whilst the net production of *Spartina* is generally high, the levels reported even from one latitude are rather variable. A major factor in this variability is the tidal range with the net production increasing as the tidal range increases, due apparently to increased availability of the nutrient nitrogen with increased tidal range. Whilst *Spartina*-dominated salt marsh estuaries certainly support coastal ecosystems through their exceedingly high productivity and the subsequent export of detritus, many of the results and conclusions are however as varied as the sites selected for study, and great care should be exercised in applying the results from one estuary to another, which may have different current patterns and topography.

Teal's (1962) study of energy flow in a salt marsh ecosystem in Georgia was one of the first studies to present a complete energy flow for any ecosystem, and he showed that the salt marsh under study received 600 000 kcal m^{-2} yr^{-1} of sunlight, of which 8295 kcal m^{-2} yr^{-1} became net primary production within the salt marsh. 20% of this net primary production was due to benthic algae, with 80% of the net primary production due to *Spartina* grass. The algae were utilised by consumer animals directly, but most of the *Spartina* became detritus and was subject to decomposition by bacteria, with much of the *Spartina* production dissipated as bacterial respiration. A small amount of the *Spartina* production was also assimilated directly by herbivorous insects (Figure 2.3).

Nitrogen is a key nutrient in the productivity of coastal ecosystems, and salt marshes which receive increased amounts of nitrogen show increased rates of primary production. The nitrogen budget of a *Spartina*-dominated salt marsh on the Atlantic coast of the USA has been investigated in detail, where it was shown that increased nitrogen supply not only increased the

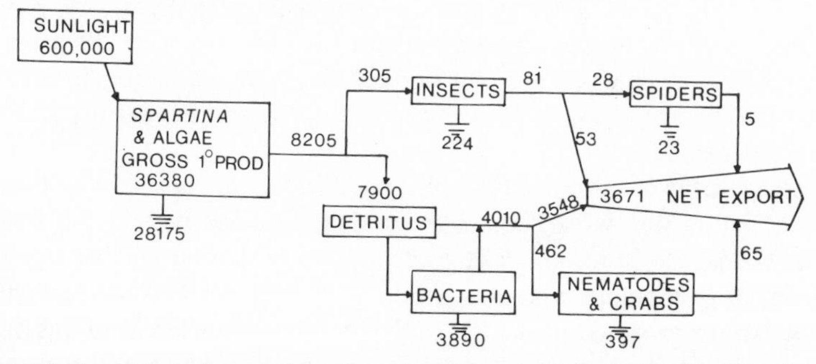

Figure 2.3 Energy-flow diagram for a Georgia salt marsh. Units are kcal m^{-2} yr^{-1}. (After Teal, 1962.)

Table 2.3 Nitrogen budget for Great Sippewissett Marsh. Adapted from Valiela and Teal (1979). Units are kg yr^{-1} for the entire 0.48 km^2 marsh.

Process	*Input*	*Output*	*Net exchange*
Precipitation	380		+ 380
Groundwater flow	6120		+ 6120
Nitrogen fixation (principally bacteria)	3280		+ 3280
Tidal water exchange	26200	31600	− 5350
Denitrification (Total loss of N gas)		6940	− 6940
Sedimentation (0.15 cm yr^{-1})		1295	− 1295
Volatilisation of ammonia		17	− 17
Shellfish harvest		9	− 9
Bird faeces	9		+ 9
Totals	35990	39860	− 3870

productivity of the plants, but also led to increased biomass in the detritus-feeding invertebrates dependent on the salt marsh (Table 2.3). Over a 2-year period groundwater flow from small underground springs, rainfall, tidal exchange, and the amounts of ammonium, nitrate, nitrite, dissolved organic nitrogen and particulate nitrogen within each were measured. Also quantified were the fixation of nitrogen by both free-living bacteria, bacteria associated with the roots of marsh plants, and by algae. Measurements were made of the loss of nitrogen from the system due to sedimentation, denitrification and harvesting of shellfish. Finally, account was made of the input of nitrogen from bird faeces. A remarkably good agreement between the measured input and the output of nitrogen is seen in Table 2.3; the 11% difference is small considering the many possible sources of error in the calculation, and in general it seems that this ecosystem is in balance. As can be seen most of the nitrogen budget is controlled by the physical factors of the tide, supplemented especially by groundwater flow.

Considerable changes do take place within the salt marsh. For example 64% of the nitrate which enters the marsh is intercepted, and ultimately leaves the marsh in the form of particulate ammonium and nitrogen. An amount of particulate organic matter equivalent to about 40% of the net annual above-ground production of the marsh is exported from this marsh, providing a rich food supply for the detritus feeders. The highly productive salt marsh studied has achieved a balanced steady state, which supports the estuarine ecosystem mainly as a source of particulate organic matter, and as a means of converting and recycling nitrogen.

Relatively little of the Spartina is consumed directly by animals, and instead the net primary production of *Spartina* mostly reaches the estuarine ecosystem in the form of fragments broken off the grass. These fragments form the basis for detritus, as they are progressively decomposed by bacteria. By trapping the detritus in the tidal creeks of salt marshes, it has been found that periodic storms are mainly responsible for the export of large quantities of detritus from salt marshes. After one storm over 2000 kg of detritus was exported in 5 hours from a 0.36 km^2 salt marsh. However, not all American estuaries receive such large quantities of detritus from *Spartina*. In the stable salt marshes of the Patuxent river estuary less than 1% of the *Spartina* production reaches the estuary as detritus. This is in great contrast to the 20–45% reported for other estuaries, which is attributed to greater degrees of tidal flooding elsewhere. The sources of detritus and its breakdown by bacteria will be discussed in section 2.5.

Dissolved organic carbon (DOC) is also released from the leaves of *Spartina alterniflora* into the estuarine water which rhythmically inundates the salt marshes. It has been calculated that the DOC released from *Spartina* is 61 kg C ha^{-1} yr^{-1}. Although this represents only a few per cent of the total production, the DOC can be readily metabolised by the microbial populations in the water and thus become efficiently available to consumer animals.

Within British and other north European estuaries the typical salt marshes are found in the region above the point of the lowest neap high tide. In this region they are not covered by the tide every day, but are covered periodically by spring high tides. Salt marshes thus occur in the intertidal area between the highest point of the lowest tide and the highest point of the highest tide, and the plants that occur there must be able to tolerate being covered occasionally by saline estuarine water. Salt marshes display a clear zonation, or successional sequence, from low to high elevations. The plant most typical of the outer, or shore, end of the salt marsh is *Salicornia* (glasswort or marsh samphire). The classical sequence is then *Glyceria maritima, Suaeda maritima*, or *Aster tripolium*, above these are *Limonium vulgare* (sea lavender), then *Armeria maritima* (sea pink), followed by *Atriplex* species, and *Festuca rubra* and *Juncus maritimus* towards the top of the salt marsh. For more details of the flora of salt marshes the reader is referred to Long and Mason (1983).

Salt marshes are thus natural eutrophic systems in which there is usually an abundance of essential elements. Salt marsh ecosystems are often systems of high gross and net productivity with normally an excess of nutrients and no dominant herbivore. Having both high biomasses and energy transfers their net productivity may exceed 1000 g dry wt m^{-2}. Table 2.2 lists some of the estimates of annual mean net productivity of salt marshes which have been made.

2.3 Intertidal plants

Seagrasses are true flowering plants, and several seagrass species inhabit estuaries, including *Thallasia, Posidonia* and *Cymodocea* in warm and tropical waters, and *Zostera, Ruppia, Potamogeton* and *Zannichellia* in temperate areas. Seagrass ecosystems, which can range in size from clumps of a few plants to large areas known as meadows, have been reviewed by McRoy and Helfferich (1977). Seagrasses, like other estuarine organisms, show greatest species diversity at the seaward and freshwater ends of estuaries, and reduced diversity within the central parts of an estuary. The eel-grass or widgeon grass, *Zostera* spp. is the commonest seagrass on the intertidal estuarine flats in many temperate estuaries growing on sandy and muddy substrata, and occurring subtidally down to 11 m depth. The annual net production is about twice the maximum biomass and ranges from 58–330 g C $m^{-2}\,yr^{-1}$, and exceptionally up to 1500 g C $m^{-2}\,yr^{-1}$. *Zostera* forms substantial communities within the Baltic Sea and the coastal estuaries of Denmark. In a study of the *Zostera* community off southern Finland, at a salinity of 6‰, it has been shown that living *Zostera*, decaying *Zostera*, and the seaweed *Fucus vesiculosus 'mf.nana'* each composed about one-third of the plant biomass. The total plant biomass was 12.5–64.4 g wet wt m^{-2}, which is low in comparison with other brackish-water plant communities, such as salt marshes.

Zostera meadows (Figure 2.4) have been extensively studied in the Isefjord, Denmark by Rasmussen (1973) who describes how much of the *Zostera* in the Isefjord, as elsewhere was destroyed in 1931–2 due, in his opinion, to an increase in water temperature throughout the 1920s, with a series of warm summers, and mild winters. Much of the *Zostera* had recovered by the 1950s with full growth in many areas, although others remain without eel-grass. In Denmark alone the net annual production of *Zostera* has been estimated at 8 million tonnes, with 6860 km^2 covered by *Zostera*. The Isefjord population shows a clear seasonal growth pattern, with peak biomass in August, witn a biomass of 443 g dry wt m^{-2}. The production of *Zostera* for the year was 856 g dry wt $m^{-2}\,yr^{-1}$ for leaf production, supplemented by 241 g dry wt $m^{-2}\,yr^{-1}$ for the below-ground

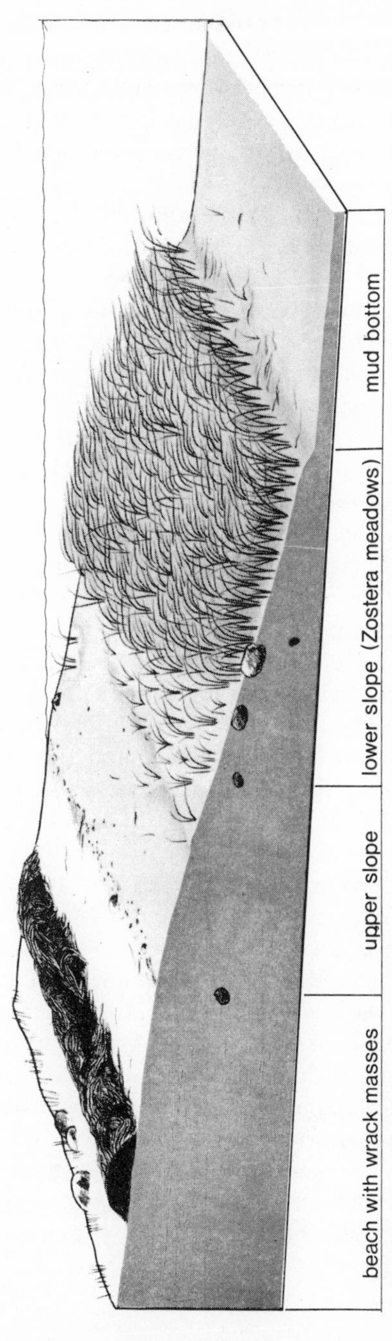

Figure 2.4 A typical eelgrass (*Zostera*) community of Danish fjords and land-locked brackish waters, in an area not subject to destruction. (From Rasmussen, 1973.)

production rhizomes. Of the leaf production, it should be noted that 69% (589 g) could be attributed to the cumulative losses of old leaves which pass out from the *Zostera* beds to become detritus in the ecosystem. Expansion of eel-grass beds has been reported in the Fraser river estuary, Canada, following the construction of a causeway which had the effect of improving water clarity.

In warmer waters, such as in Florida or Puerto Rico, *Thalassia* becomes the main seagrass, with biomasses of 20–8100 g dry wt m^{-2} and productivity values of 100–825 g C m^{-2} yr^{-1}. These high productivity values will often be supplemented by 20–30% by epiphytic plants, that is smaller plants growing attached to the *Thalassia*. Again, as for *Zostera*, the energy may be utilised by animals, not so much by grazing the seagrass, but rather through the detritus route.

The production of seaweeds (macroalgae) such as *Fucus* and *Ascophyllum* can be high on marine rocky shores. However in estuaries, populations of seaweeds tend to cover a very small proportion of the total area, being confined to rocky outcrops, quays and piers. The seaweed *Fucus ceranoides* is confined to estuaries, in contrast to other *Fucus* species which tend to occur only on fully marine coasts. In a comparison of the estuarine *Fucus ceranoides* with the marine *Fucus vesiculosus* it has been found that the distribution of the species are limited by salinity, with low salinity unfavourable for *Fucus vesiculosus* and high salinity unfavourable for *Fucus ceranoides*. The seaweeds of Florida estuaries, which are tolerant to a wide range of temperature, light and salinity and even short exposures of fresh water, can continue to photosynthesise whilst both covered and exposed by the tide. Within the Asko area of the Baltic Sea, the net production of the *Fucus* is 20% of the phytoplankton production in the same area, but both production values are low in comparison to the production values in salt marshes nearby.

The mudflats of estuaries, which receive high nutrient (especially nitrogen) inputs from inland areas, for example the Eden, the Ythan or Chichester harbour in the UK, may become covered in profuse growths of the green alga *Enteromorpha* (mainly *E. prolifera*) which develop as mats during the summer season, and decline in the autumn. At the end of the growing season large populations of heterotrophic bacteria, and subsequently de-nitrifying bacteria, develop on the rotting algae. The *Enteromorpha* can be the main means of accumulating nitrogen from the waters which flow into the estuary, and as the algal mats decay the nitrogen is made available to other parts of the ecosystem. The mats may also smother the animals living within the mudflats, and as the mats decay they may utilise much of the available oxygen, to the detriment of the animals.

Large populations of diatoms and other microalgae, known as *microphytobenthos* or *epibenthic algae*, occur in the upper 1 cm of mudflats, although living diatoms can be found down to 18 cm due to diurnal vertical migration within the sediment. The richest populations of microalgae have generally been found on the lowest parts of the intertidal areas. In contrast to phytoplankton which typically has pronounced seasonal fluctuations in number and biomass, some authors have found no seasonal fluctuations in the benthic microalgae due to the continuous regeneration of nutrients by bacteria within the sediment. In the Wadden Sea, Netherlands (Figure 2.5) and Lynher estuary, UK, however, a clear seasonal pattern to the production of the microphytobenthos appears to be closely linked to temperature variations. The microphytobenthos can have a significant role to play in the mudflat estuarine ecosystem, with values of net production of 30–300 g C m^{-2} yr^{-1} (Table 2.4). Within the Lynher estuary, the primary production of epibenthic algae can be compared with the phytoplankton production in the overlying water. The annual net production for the benthic algae, at 143 g C m^{-2} yr^{-1} being almost double the value of 81.7 g C m^{-2} yr^{-1} for the water column. Much of the epibenthic algae appears to be utilised by bacterial populations within the mudflat surface and these, together with the algae, are utilised by the consumer animals. Benthic microalgae have a valuable role to play in the

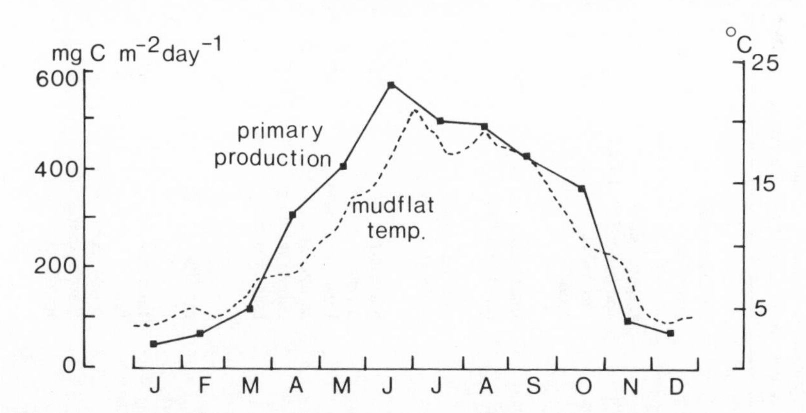

Figure 2.5 Average primary production of microphytobenthos on a tidal flat in the Wadden Sea, near the island of Texel, The Netherlands. Production is expressed as mg C m^{-2} day^{-1}, and the temperature of the upper sediment layer of the same mudflat is also shown. (After Cadee, in De Wilde, 1975.)

Table 2.4 Comparison of annual production rates of microphytobenthos in intertidal and shallow coastal sediments. Upper part: ^{14}C-values; lower part: O_2-values. (After Colijn and de Jonge, 1984.)

Locality (Lat)	*Sediment type, depth (m)*	*Method*	*Annual production ($g\,C\,m^{-2}$)*	*Production rate ($mg\,C\,m^{-2}\,h^{-1}$) range; mean*	*Chlorophyll ($mg\,m^{-3}$)*	*Dominating microphytes*
Danish fjords (55°N)	Littoral sand and mud, 0.2–1.8 m	^{14}C	116	25–90; 60	n.d.	Pennate diatoms
Ythan estuary Scotland (57°N)	Intertidal muds	^{14}C	31	4–26; 10	25–34 $\mu g\,g^{-1}$ dry sediment	Motile pennate diatoms
South New England, USA (41°N)	Intertidal mixed sediment	^{14}C	81	8.2–30.8; 20.1	100	Pennate diatoms dinoflat., filamentous algae
Madagascar (13°S)	Marine sands, 5–60 m	^{14}C	150 (5 m) 66 (mean)	9.22	38.78	Cyanophyceae, diatoms, symbiotic dinoflag
Falmouth Bay, USA (41°N)	Salt marsh muds	^{14}C	105.5 ± 12.5	5–80	n.d.	n.d.
Chuchi Sea, USA (71°N)	Fine muds and sands, 5 m	^{14}C	5	0.5–57	40–320	Diatoms and Euglenophyceae
Wadden Sea, Netherlands (53°N)	Intertidal flats (sandy-mud)	^{14}C	100 ± 40	50–100 (winter) 100–1100 (summer)	7.1 $\mu g\,g^{-1}$ dry sediment 100; 40–400	Attached and suspendable diatoms
River Lynher estuary, SW England (50°N)	Intertidal mudflats	^{14}C	143	5–115	30–80 $\mu g\,g^{-1}$ dry sediment	n.d.
Balgzand, Wadden Sea (53°N)	Intertidal flats, 4 transects	^{14}C	85 (29–188) (15 stations)	0–900 (d^{-1})	3–13 $\mu g\,g^{-1}$ dry sediment	Diatoms
Bolsa Bay, USA (34°N)	Barren estuarine mudflats	^{14}C	115–246	26–59 (4 stations)	185–385 (annual mean)	Motile and non-motile diatoms, bluegreens, dinoflagellates

Ems-Dollart estuary (53°N)	Intertidal mudflats	^{14}C	62–276 (6 stations)	1–120; 37.0	3–184 (annual mean)	Diatoms, occ. Euglenophyceae bluegreens
Mms-Dollart estuary (53°N)	Intertidal mudflats	O_2	69–314 (6 stations)	0–1900 mg C $m^{-2} d^{-1}$	n.d.	Diatoms, occ. Euglenophyceae bluegreens
False Bay, USA (48°N)	Intertidal sandflats	O_2	143–266 (3 stations)	0–100	30–70 μg^{-1} dry sediment	Diatoms (*Navicula* spp.)
Georgia salt marshes, USA (31°N)	Intertidal mudflats	O_2	200	5–140	n.d.	Pennate diatoms flagellates, bluegreens, dinoflagel
Bay of Fundy, Canada (45°N)	Intertidal flats	O_2	47–83	10–800	10–500	Microalgae

n.d. = not determined.

formation and maintenance of an oxygenated zone on the surface of intertidal estuarine sediments. Along with the physical forces of the tide, microalgae may be the main source of oxygen for the sediment surface through the process of photosynthesis.

2.4 Phytoplankton

The phytoplankton is an integral part of the estuarine ecosystem; however, it does not have such a dominant role as for example the phytoplankton in marine ecosystems or freshwater lakes. Despite the abundance of nutrients in estuaries, other factors may limit the production of estuarine phytoplankton.

The photosynthesis and respiration of phytoplankton has been measured in a 400 km^2 system of estuaries near Beaufort, North Carolina. The net production was 52.5 g C m^{-2} yr^{-1}. Whilst the daily rate of production of phytoplankton could be quite high, the annual rate is relatively low, which might be due to two factors, shallowness and turbidity. The shallow nature of the estuaries studied, which is typical of most estuaries, meant that the mean depth of the water, at 1.18 m, was 1.7 m less than the optimum depth for producing maximum net photosynthesis. The penetration of light in estuarine waters is severely limited by the turbidity of the water, due to suspended sediments and particulate organic matter, which will again limit the production of the phytoplankton. However these two factors do not entirely explain the low rate of net production, and it is considered that the key factor in the estuaries studied might have been the lack of a spring bloom, the absence of which is attributed to the shallow nature of the water and the continual mixing which leads to the lack of any stratification.

Within the Lower Hudson estuary, USA, dissolved inorganic nutrients are high throughout the year, but large blooms of phytoplankton do not occur despite this availability of nutrients. Malone (1977) suggests that this is due to the flushing rate of the estuary, whereby the populations of phytoplankton are carried out to sea before their growth rates permit the development of phytoplankton blooms. Those peaks of phytoplankton that do occur are related to the incursion of marine water carrying plankton into the estuary. Similarly, the production of the phytoplankton of the Zuari and Mandovi estuaries in Goa, India has been found to increase at high tide due to the incursion of marine water containing rich phytoplankton.

The plankton community of the southwest Bothnian Sea (part of the

Baltic Sea) has a large phytoplankton biomass in the spring, but a small biomass for the rest of the year in this low salinity sea, where conditions are normally 5–6‰. The spring bloom is dominated by diatoms and dinoflagellates, especially *Thalassiosira* species. The biomass of the total phytoplankton was between 0.464 and 0.394 g C m^{-2}, averaged over two years (equal to 0.928–0.788 g dry wt m^{-2}). Feeding on this phytoplankton was 0.054 g C m^{-2} of protozoa, rotifers and nauplii, and up to 0.482 g C m^{-2} of larger zooplankton, especially the copepods *Acartia bifilosa* and *Eurytemora affinis*. Production was not measured in this study, but comparisons with other Baltic regions suggest a value of 48–94 g C m^{-2} yr^{-1} to be appropriate. This value is lower than production values for coastal phytoplankton, and much lower than values for salt marsh production (Table 2.2), but the high productivity in relation to the biomass is pronounced, and the phytoplankton is for many animals a richer source of food than plant fragments prior to decomposition. The ciliate protozoa, with their rapid turnover time (2 days) may often be the main consumer of phytoplankton in brackish waters, consuming more than the heavier, but slower-growing, copepods of the zooplankton.

As freshwater phytoplankton are carried into estuaries, they are killed at salinities greater than 8‰, and it has been hypothesised that the oxygen minima observed at the freshwater–brackish-water interface may be due to mass mortality of these algae. Studies in the Tamar estuary suggest that the oxygen minima are more closely related to the position of the turbidity maximum, but this increase in turbidity will certainly inhibit the algae.

Research workers studying different estuaries have come to widely different conclusions regarding the role of phytoplankton, some claiming that primary production of phytoplankton is insignificant, whilst others regard phytoplankton production as being of central importance to the estuarine ecosystem, responsible for approximately 85% of total ecosystem primary production. Although nutrients appear to be frequently available for the production of large quantities of phytoplankton, maximal production is apparently rarely achieved due to three factors. Firstly, turbidity can limit the penetration of light, secondly, the shallowness of many estuaries means that blooms may not develop, and thirdly, the growth rate of the phytoplankton may be less than the flushing rate of the estuary.

Table 2.5 summarises the data on primary production from six estuaries, where the various components have been measured, and allows us to examine the relative contributions of the various producers. It must first be emphasised how variable the total production is, with total production ranging from 63.6 to 1445 g C m^{-2} yr^{-1}, and the examples given are from

Table 2.5 Net primary production of particulate material in various estuaries, expressed as percentage of total production. Data from Knox (1986).

	Phyto-plankton	*Zostera*	*Epiphytes*	*Spartina*	*Mangrove*	*Epibenthic algae*	*Mangrove (Avicennia)*	*Total production* $g\,C\,m^{-2}\,yr^{-1}$.
	As % total net primary production							
Beaufort, N. Carolina	43.3	38.0	8.5	9.8	0	Not measured	0	152.6
Flax Pond, New York	2.2	0	3.7	74.7	20.5	5.6	0	535.0
Sapelo Island, Georgia	6	0	0	84	0	10	0	1445
Nanaimo estuary, Brit. Columbia	11.8	42.1	0	0	6.6	39.5	0	63.6
Barataria Bay, Louisiana	18.9	0	0	69.0	1.1	22.2	0	890
Waitemata Harbour, New Zealand	25.3	0	0	0	0.3	30.8	31.4	473.5

various latitudes with conditions ranging from mangrove and *Spartina*-dominated estuaries, through to estuaries dominated by bare mudflats. Phytoplankton production contributed between 2.2 and 43.3%, whilst epiphytes were less than 8.5% where studied, and macroalgae (mainly fucoids) also contributed little, except in Flax Pond, where they supplied 20.5% of net production. In the estuaries with bare mudflats, epibenthic algae contributed over 30% of production, but in those dominated by *Spartina* it was much less. When present, *Spartina* supplied up to 84% of total primary production. From these various studies, it must be concluded that each estuarine ecosystem has its own characteristics, with a unique mix of primary producers. As far as the primary consumers are concerned, the mix of primary producers may not be very important, if most energy is consumed in the form of detritus, and it may be the supply of detritus derived from the breakdown of the primary producers which is the feature of most importance to the success of the primary consumers.

2.5 Detritus

Detritus has already been defined as all types of biogenic material in various stages of microbial decomposition. Much of this biogenic material may be fragments of plants, such as *Spartina*. In estuaries without large salt marshes the main sources of detritus are fragments of dead plants and animals from the sea, from rivers, or from the estuary itself, as well as the faeces, and other remains of the estuarine animals. All the types of primary production described in the preceding sections of this chapter can supply material which becomes detritus, and it is clear from many studies that most primary production in estuaries is not consumed directly by herbivores, but rather is converted into detritus before consumption by detritivores.

Spartina and other plant detritus is relatively indigestible to the consumer animals and thus much of the flux of organic matter to detritivores must involve the conversion of the particulate detritus to soluble compounds and their assimilation by micro-organisms, which can then be consumed by detritivores. The nutritive value of *Spartina* increases as the detrital fragments become enriched with microbial populations. In Figure 2.6 it can be seen that living *Spartina* has a content of 10% protein. Dead leaves entering the water have about 6% protein, but as the plant fragments become smaller the protein content increases to 24%. Thus the detritus which is rich in protein may be a better food source for animals

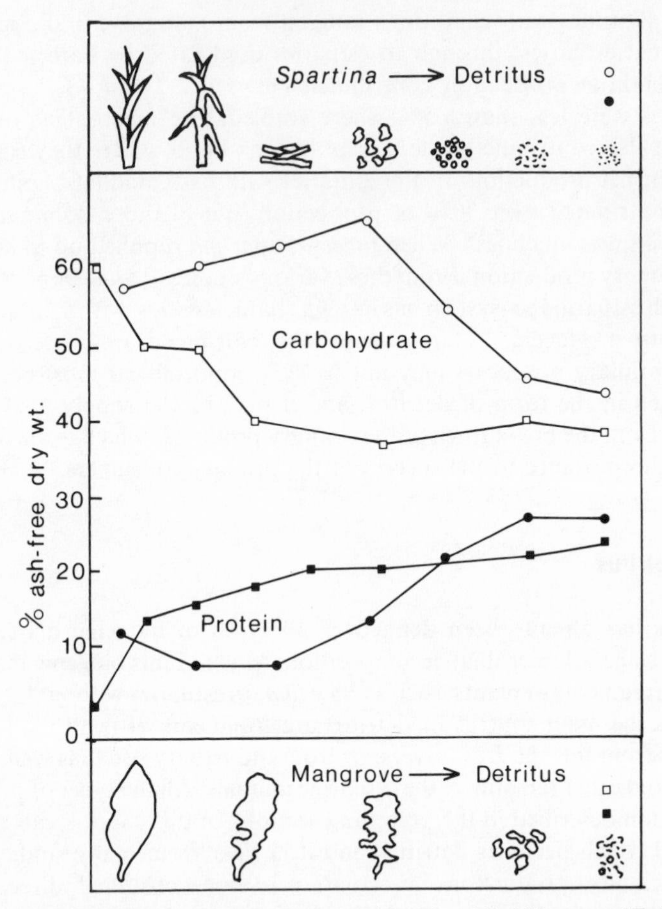

Figure 2.6 Change in the constituents of *Spartina* (○, ●) and Red Mangrove (□, ■) leaves during conversion from living plant material to fine detritus fragments, as shown pictorially. (After Odum and de la Cruz, 1967, and Heald, 1969.)

than the grass tissue that formed the basis for the particulate matter. Similar results have been described for leaves of the tropical estuarine salt-marsh plant, red mangrove, with 6.1% protein in leaves on the tree, 3.1% protein at leaf fall, and 22% protein after decomposition in estuarine water for 12 months. The main decomposer of plant material in seas and estuaries are bacteria, as shown in Figure 2.7.

The detritus, composed of the decaying remains of plant primary production, and microbes, has a valuable role in stabilising the estuarine

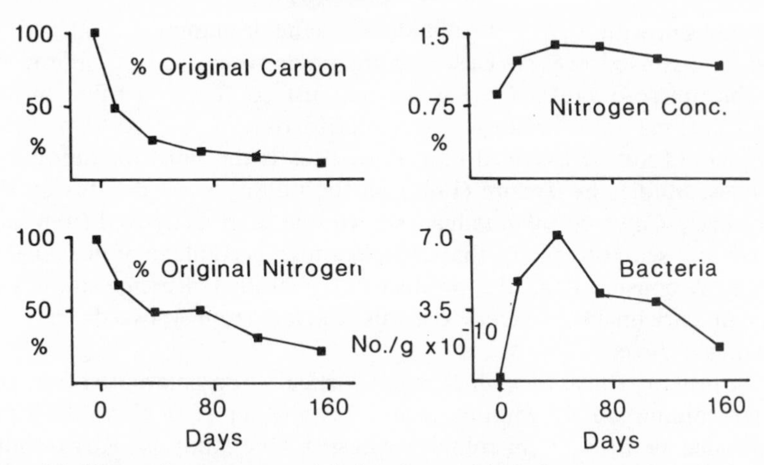

Figure 2.7 Changes in percentage original carbon, original nitrogen, and total nitrogen concentration, and numbers of bacteria, during the decomposition of submerged leaves of the mangrove *Avicennia marina.* (After Robertson, 1988.)

ecosystem by levelling out the seasonal variations in primary production, ensuring a year-round food supply, and securing the reabsorption of dissolved nutrients. The role of micro-organisms in the process of the breakdown of plant material in estuaries may be compared to the role of micro-organisms in the guts of terrestrial herbivores, as the bacteria living on particulate or dissolved organic matter in both cases make the primary production more readily available for animal consumption.

When bottom-dwelling animals consume detritus, it appears that they consume the bacteria and other microbes, but reject the plant tissues. In the process they may shred the plant material into finer fragments, which will provide a larger surface area for micro-organisms, and so accelerate the processes of decay. The activities of animals which consume the entire sediment, lead to a continual mixing of the organic and inorganic particles in the sediment, in a process known as bioturbation. This will tend to distribute detrital material throughout the surface layers of the sediment, and so enable material which has settled on the surface of the sediment to organically enrich the sediment to a depth of several centimetres. It has been calculated that the biomass of bacteria within estuarine sediments may be of the same order of magnitude as the biomass of the animals in the sediment. Apart from the immediate surface layer, estuarine sediments tend to be anaerobic, as the bacteria and other micro-organisms consume all the available oxygen. Much detritus therefore undergoes anaerobic metabolism, with hydrogen sulphide, methane or ammonia produced, as well as dissolved organic carbon compounds which can be utilised by aerobic micro-organisms living on the surface. These aerobic micro-organisms may also be consumed by detritivores.

It should not be assumed that all detritus is the same for detritivore animals. Studies by Tenore (1981) on the utilisation of detritus by the polychaete *Capitella capitata* have shown that detritus derived from salt-marsh and sea-grass plants that contain a high percentage of unavailable energy is consumed as the products of microbial decomposition and protein enrichment, whereas detritus derived from seaweeds may be consumed directly.

The primary consumers which ingest particulate organic matter from the water column do so without regard to whether it is phytoplankton, suspended detritus or microbial organisms. One study by Kirby-Smith (1976) attempted to segregate these components, and came to the conclusion that the bay scallop, *Argopecten irradians*, ingests 20% phytoplankton and 80% detritus and bacteria, emphasising the much greater availability of detritus within the estuarine ecosystem, even though the growth rate of the scallops would have been higher on a diet of phytoplankton alone. The activities of these suspension-feeding bivalves may be profound. In South San Francisco Bay estuary, for example, the suspension-feeding bivalves filter a volume equivalent to the total volume of the area each day, and this grazing may be the primary mechanism controlling phytoplankton biomass. In Chesapeake Bay, USA, the detritus averages 77% of the total organic particles in the water column, and the phytoplankton 23%. The numbers of detritus particles in the water showed little seasonal variation, whereas the phytoplankton in this area showed considerable seasonal variation. In the Dollard estuary, a gradient has been observed in the percentage of 'living' carbon in the particulate organic matter in the water column, with low values in the inner estuary (2.3–15.4%), and higher values (5–57%) in the outer estuary. The proportion of living phytoplankton increased in the summer, especially at the outer estuary, but otherwise detritus formed the majority of the particulate organic carbon in the estuary (Figure 2.8).

Williams (1981) has estimated the global organic inputs into estuaries (Table 2.6), and shows that the major sources are primary production from both wetlands (salt marshes) and planktonic and intertidal algae, along with organic matter carried into the estuary from rivers. To these values must be added man's discharges of sewage, oil products, food products and wood pulp. There must also be a quantity of organic matter entering the

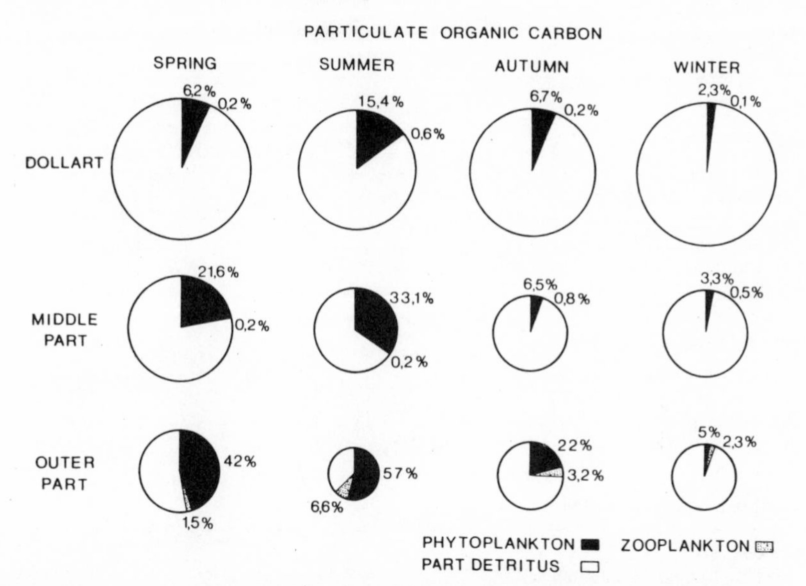

Figure 2.8 Mean concentration of total particulate organic carbon (area of the circle), mean concentration of phytoplankton (in black), zooplankton carbon (stippled), and particulate detritus (in white). Numbers indicate phytoplankton and zooplankton carbon as percentage of the total particulate organic carbon. Data given for the Dollard estuary, and middle and outer parts of the Ems-Dollard estuary (see Figure 1.9) in spring, summer, autumn and winter, for 1978–80. (From Laane, 1982.)

estuary from the sea, but Williams has not quantified this. All these sources of organic matter are utilised by micro-organisms within the estuaries of the world, to become detritus.

Along with the particulate organic matter that forms detritus in estuaries, there may be considerable quantities of dissolved organic matter present, derived from plant exudation, animal excretion and from the products of decomposition. The capacity for the uptake of dissolved organic matter by animals is widespread, but despite this it seems likely that estuarine animals get the vast majority of their food from particulate organic matter. The dissolved organic matter will mainly be metabolised by bacteria, and some estimates show that bacterial production utilising dissolved amino acids can amount to 10% of algal production. The bacteria, consuming the dissolved organic matter, themselves become part of the particulate matter in the estuary.

Table 2.6 Provisional estimates of global organic inputs into estuaries. (From Williams, 1981.)

Source	*Annual flux*	
	log_{10} g C per yr	*g C per yr*
INTERNAL SOURCES		
Wetlands photosynthetic production (estimate from Woodwell *et al.*, 1973)	14.6	3.8×10^{14}
Planktonic and intertidal photosynthetic production	14.1	1.4×10^{14}
EXTERNAL SOURCES		
Rivers (estimated median value)	14.3	2×10^{14}
Sewage (*maximum* estimate)	13.0	1×10^{13}
Oil and oil products (*maximum* estimate)	13.4	2.5×10^{13}
Food and food products (*maximum* estimate)	13.7	5×10^{13}
Wood and wood pulp (*maximum* estimate)	13.0	1×10^{13}
Total (*maximum* estimate)		8×10^{14}
CARBON FLUX IN OTHER MAJOR ECOSYSTEMS		
Terrestrial production (estimate from Whittaker and Likens, 1975)	16.8	69×10^{15}
Marine production (estimate from Whittaker and Likens, 1975)	16.3	22×10^{15}

2.6 Summation of plant and microbial production in estuaries

The various components of primary and microbial production can be combined in an attempt to understand a particular estuary, and to try and explain the high productivity of estuaries in general.

Measurements have been made in the Dollard estuary on the Dutch–German border in an attempt to quantify all sources of organic input to this estuary. Eighty percent of the estuary is composed of tidal sand/mudflats. The total amounts of organic carbon entering and leaving the Dollard estuary are shown in Table 2.7. This preliminary attempt at a carbon budget has clearly revealed a large discrepancy between the measured inputs and outputs, which is believed to be mainly due to unquantified export of dissolved carbon from the estuary. Nevertheless, several valuable points emerge from this study. Firstly, the main sources

Table 2.7 Organic carbon budget for the Dollard estuary. Adapted from van Es (1977). Units are $\times 10^6$ kg C yr^{-1} for the entire area of almost 100 km^2.

Import/production		*Export/utilisation of organic carbon*	
Particulate C from North Sea + River Ems	37.1	Dissolved C to North Sea	?
From potato flour mill	33.0	Utilisation in water	7.2
From salt marshes	0.5	Utilisation in sediment	18.2
Primary production phytoplankton	0.7	Buried in sediment	9.9
Primary production of benthic algae	9.3	Bird feeding	0.26
	80.6		35.56

(75%) of carbon are outside the estuary in the river, the sea, and an industrial plant (potato flour mill) which discharges effluent. Secondly, due to the turbidity of the water the primary production from phytoplankton is only 7.5% of the primary production from benthic algae such as diatoms and blue-green algae. Furthermore, the total primary production of 10×10^6 kg C yr^{-1} is considerably less than the carbon consumed, or utilised in the water and sediments (25.4×10^6 kg C yr^{-1}). It should be noted that the value for the utilisation of carbon includes mainly the amount of carbon utilised by the detritus feeders within the estuary. The first step in this utilisation of carbon is the breakdown of detritus by bacteria, and detritus feeders principally feed in the bacterial populations.

The study of the Dollard estuary clearly shows that primary production within an estuary is inadequate to support the large number of detritus feeders inhabiting the mudflats, and the detritus feeders must rely on the importation of organic debris from outside the estuary. What primary production does take place is due overwhelmingly to the benthic algae, rather than the phytoplankton whose production is inhibited by the turbidity of the water. Thus the basic biological processes creating energy for the primary consumers in this estuary are concentrated on the mud surface with the primary production of the benthic algae, and the transformation of organic debris into more digestible material by bacteria. The high productivity of the Dollard, and many similar estuaries, is thus seen to be due to the position of the estuary as a collecting area for organic matter, supplemented by the primary production of the benthic algae on the intertidal flats. The high productivity of this estuarine ecosystem is

Table 2.8 Food sources for the Grevelingen estuary, The Netherlands, expressed as the import and production of particulate organic carbon. (Adapted from Wolff, 1977.)

Food source	$g\,C\,m^{-2}\,yr^{-1}$
From salt marshes	0.3–7
Production *Zostera*	5–30
Production microbenthic algae	25–37
Production phytoplankton	130
Land run-off	2
Import from North Sea	155–255
Total of sources	317–451

because it is subsidised by the transfer of energy from other ecosystems.

The Grevelingen estuary, in the Netherlands, was studied intensively prior to the implementation of the Delta Barrage scheme, which is described in Chapter 5. This estuary covered 140 km^2, of which 81 km^2 was covered at all times, 55 km^2 was intertidal sand and mudflats, and 4 km^2 was salt marsh. A detailed food budget for the period before it was dammed is shown in Table 2.8. It is assumed that 10–45% of the above-ground production of the salt marsh was exported from the salt marsh as detritus, but due to the small area of marsh the net contribution from the marsh is small. The eel-grass, *Zostera* covers up to 12 km^2. 10% of the annual production is directly consumed by the birds, wigeon (*Anas penelope*) and brent geese (*Branta bernicla*), leaving 5–30 $g\,C\,m^{-2}\,yr^{-1}$ available for detritus. The levels of primary production are dominated by the production of phytoplankton, supplemented by benthic microalgae. This is the reverse of the situation in the Dollard estuary, and is due to the less turbid waters in the Grevelingen, coupled with the smaller proportion of intertidal area. The total contributions from all sources within the estuary to the carbon budget is however exceeded by the material carried in on each tide from the adjacent North Sea. It will be shown in the next chapter that the Grevelingen estuary supported a primary consumer benthic community with a production of 57.4 g ash-free dry wt $m^{-2}\,yr^{-1}$ (= 30.3 g C). If the value of 30.3 g C is compared to the values of 317–451 g C from Table 2.5, it can be seen that 6.7–9.5% of the food sources becomes macrofaunal primary consumer production.

Carbon budgets have been calculated for the intertidal areas of the Wadden Sea by Postma (1988), as shown in Table 2.9, which again reveal

Table 2.9 Possible carbon budget for a Wadden Sea tidal flat. (From Postma, 1988.)

		1970	*1980*
GAINS $g C m^{-1} yr^{-1}$	Benthic algae	100[1]	200[1]
	Phytoplankton	70	115[2]
	Total primary production	170	315
	Import	240	300[3]
	Total gains	410	615
LOSSES	Diss. Org. C	60	90
	Zooplankton + bacteria	40	60?
	Meiofauna	14	14
	Macrobenthos	60	100
	Sediment Aerobic	118?	225?
	Sediment Anaerobic	118?	225?
	Total losses	410	615

[1] Assuming no losses to creeks.
[2] Via chlorophyll increase.
[3] Lake Ijssel added.

the importance of the importation of material from the sea and from river sources (mainly Lake Ijssel). If the budgets for 1970 and 1980 are compared, the extent of eutrophication that took place in the intervening period can be assessed. The budgets show that primary production, due to benthic algae and phytoplankton both increased, as well as the importation of carbon. Most of the faunal components increased their comsumption, with the greatest increase being in aerobic and anaerobic utilisation of the carbon within the sediments. These examples given so far thus all depend in large part for their productivity on the importation of energy.

The carbon budget for Barataria Bay, Louisiana shown in Table 2.10

Table 2.10 Carbon budget for Barataria Bay, Louisiana, expressed as $g C m^{-2} yr^{-1}$, assuming that 1 g C is equal to 2 g dry organic wt. (From Wolff, 1977.)

Source		*Sink*	
From saltmarshes	297	Consumed in estuary	432
Production benthic algae	244	Exported to sea	318
Production phytoplankton	209		
Totals	750		750

Figure 2.9 The Forth estuary, eastern Scotland, UK. A typical 'European'-type estuary with large intertidal mudflat areas, bare of macrophyte vegetation. Microphytobenthos is a main primary producer in such habitats. In the foreground, the plant *Salicornia* can be seen colonising the uppermost areas of the mudflat. In the background can be seen a large coalfired power generating station (see Chapter 5).

Figure 2.10 The Great Bay estuary, New Hampshire, USA. A typical 'American'-type estuary where the macrophyte *Spartina* occupies much of the intertidal habitat. In the foreground, fragments of *Spartina* are decomposing, and ultimately supplying detritus for the ecosystem.

reveals that unlike the previous examples, it is a net exporter of energy rather than a net importer. Whilst the primary production within the estuary of phytoplankton and benthic algae is important, the largest source of energy is the supply of detritus from the *Spartina*-dominated salt marshes.

From a consideration of the energy budgets presented above, it is clear that two distinct types of estuary emerge, although there is undoubtedly a spectrum of types, with the most distinct examples at the opposite ends of the spectrum. At one extreme are the 'European-type' estuaries, such as the Dollard, which are dominated by large, relatively bare intertidal mudflats (Figure 2.9), and at the other extreme are 'American-type' estuaries which are dominated by large stands of the marsh grass *Spartina* (Figure 2.10).

In the 'European-type' estuary much of the primary production within the estuary is performed by large populations of microscopic benthic algae living on the surface of the mud (Figure 2.9), supported by phytoplankton in the water column. The extent of the primary production of the phytoplankton depends on the turbidity of the water. However in these estuaries the majority of the energy within the primary producer trophic level is derived from outside the confines of the estuary, and is in the form of organic matter which is carried into the estuary, usually from the sea, but also from land discharges of river water or sewage. The estuary is thus a net recipient of energy, and the high productivity which supports large populations of consumer animals is due to the position of the estuaries as traps for both nutrients and particulate organic matter.

In the 'American-type' estuary the primary production of benthic algae and phytoplankton is important for the productivity of the whole ecosystem, but the dominating factor is the much greater proportion of the estuary which is inhabited by rich beds of *Spartina* grass. The *Spartina* is only consumed directly by animals to a small extent, and instead they rely on the fragments of *Spartina* forming the substrate for large populations of bacteria, which form detritus, which is then ingested by the animals (Figure 2.10). However, despite high rates of consumption within the estuary, excess material remains which is carried out of the estuary to fertilise the adjacent sea.

For both types of estuary, and those intermediate between the two extremes, we can conclude that the high levels of production within estuaries are due to a plentiful supply of nutrients supporting the primary production of benthic algae, phytoplankton and salt marshes, enhanced by the import of particulate organic matter into the estuary from either the sea or the margins of the estuary, which undergoes microbial decomposition within the estuary to yield a rich food supply for the consumer animals.

CHAPTER THREE

PRIMARY CONSUMERS HERBIVORES AND DETRITIVORES

3.1 Introduction

Estuaries are rich in food sources for the primary consumer trophic level in the food web. Phytoplankton, as we have seen, is limited by turbidity but is nevertheless a rich source of food. The main food source is however the large quantities of detritus which abound in the water column and on the bottom of the estuary. The supply of food is replenished both by tides and by freshwater inflow, and the deposition of fine particulate matter and detritus in the central reaches of the estuary provides a food store which is available for virtually the whole year. Whereas food chains in temperate seas and freshwater lakes are dominated by short bursts of primary production, especially in the spring, estuaries are characterised by having food sources available for the whole year, although the food sources are richer in the spring and summer as increased temperatures accelerate all biological production.

Most of the primary consumer animals are found on the bottom of the estuary, where a rich benthic community usually develops. Zooplankton is usually present in the water column, but the strong tidal currents and river flow which flush out the estuary, coupled with the limitations imposed by turbidity, make zooplankton a less dominant feature of the estuarine food web than food webs in the sea. For both zooplankton and benthic animals, most food is in fine particulate form, whether it is live phytoplankton or the variously decomposing fragments which make up the detritus. Detritus is a rich food source for the primary consumers, and indeed it is usually accepted that bacteria-rich detritus is often a better source of food for the primary consumer than the plant tissues which formed the original material for much of the particulate organic matter.

It is possible to attempt to classify the benthos into suspension (or filter) feeders relying on fine particulate matter suspended in the water, and deposit feeders which rely on the food contained within the muddy estuarine deposits. However, the borderlines between these two groups are often unclear in the estuarine ecosystem. Among the suspension feeding members of the estuarine benthos, the most typical is perhaps the common mussel, *Mytilus edulis*, which relies on phytoplankton and small organic particles as its food source. However, its filtering process accepts food particles within a specific size range ($>2\,\mu$m) and it is unable to distinguish between phytoplankton and floating detritus of similar size. Among deposit feeders, the typical example may be the lugworm, *Arenicola marina*, which ingests large quantities of mud, and having digested off any organic material, rejects the bulk of the mud in the familiar worm cast.

However, between these clear-cut examples are many animals which can both suspension-feed and deposit-feed. The baltic tellin, *Macoma balthica*, lies buried in the mud with its siphons protruding (Figure 3.1). When food is available in the water overlying the mud, it is able to raise the incurrent siphon and draw in water and particulate matter, but when detritus particles are available on the surface of the mud surrounding its buried location, it can reach out across the mud and collect the particles by using its siphon as a 'vacuum cleaner'. Within the animal the same gills provide a size-limited filtration for the food particles, so that *Macoma* for example, can spend 10–40% of its life suspension-feeding and 60–90% in deposit-feeding.

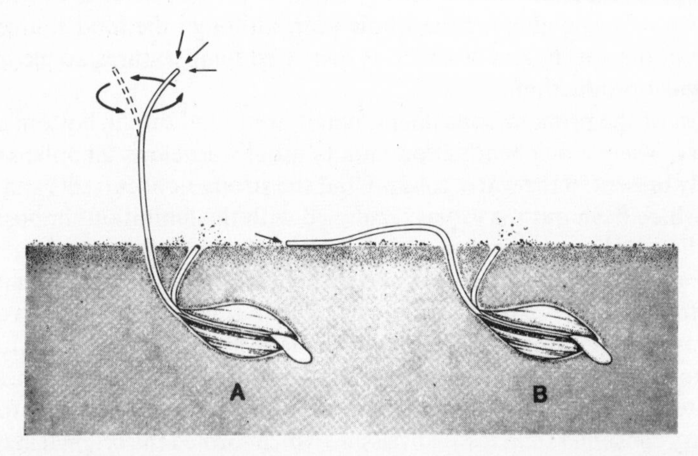

Figure 3.1 *Macoma balthica*. The left individual is seen feeding on suspended matter with the inhalent siphon stretched up in the water. The right individual is seen sucking in bottom material with the inhalent siphon. For both, the exhalent siphon rejects waste material. (From Olsen, in Rasmussen, 1973.)

It has been generally considered that the major food source for deposit feeders is the microbes attached to sediments and detritus particles, but it has been difficult to specify the actual food sources for individual animals. The food ingested can, as has been shown by Lopez and Levinton (1987), either be derived from 'microbial stripping', or from amorphous organic debris. It seems likely that both sources are required to support deposit feeders.

The food supply and the numbers of suspension-feeding benthic animals appear to fluctuate more heavily, spatially as well as temporally, than those of deposit feeders. This is because the food supply for suspension feeders, which is phytoplankton and suspended particulate organic matter, varies seasonally during the year, or even from day to day, whereas the food supply for deposit feeders, which is benthic algae or detritus and the microbes growing on it, will vary less throughout the year.

As an alternative to the classification of the benthos into suspension feeders and deposit feeders, the system of epifauna (or surface fauna) and infauna (or buried fauna) may be suggested. Again this is sometimes clear, as one can readily classify as epifauna animals on the mud surface such as mussels (*Mytilus*) on a mussel bed, barnacles (*Balanus*) and winkles (*Littorina*) on rocky outcrops, or mobile animals such as mysid shrimps (*Neomysis*) and gammarid amphipods (*Gammarus*). Among the infauna are clearly many of the estuarine worms, such as the polychaetes *Nereis, Nephthys, Arenicola* or the oligochaetes, *Tubificoides, Tubifex*. However, other common animals spend some of their time buried in the mud, and then emerge onto the surface to feed. Typical examples of these are the small gastropod snail *Hydrobia*, and the crustacean amphipod *Corophium*.

When examining estuarine benthic samples it is frequently desirable to sort them on the basis of size, with animals retained by a 500 μm (0.5 mm) sieve classified as macrobenthos, animals retained by a 40–60 μm sieve classified as meiobenthos, and animals passing a 40–60 μm sieve classified as microbenthos. However this classification can lead to the small worms such as oligochaetes, which can be numerically dominant on polluted mudflats, being neglected.

In order to examine the productivity of this important primary consumer trophic level we shall examine first the macrofaunal mud dwellers, which are mostly infaunal deposit feeders, but which may make excursions into the epifauna or utilise the suspension mode of feeding. Then we shall consider the macrofaunal surface dwellers considering both mobile and static representatives which are predominantly suspension feeders. The potentially important, but less studied, meiobenthic fraction will also be considered, as will the zooplankton. Finally the reasons for the high levels of production will be examined.

3.2 The mud dwellers

3.2.1 *Molluscs*

The bivalve *Macoma balthica* (Figure 3.1) is one of the most widespread of estuarine benthic animals. It can be a rather slow-growing animal, and in some situations such as the St Lawrence in Canada may take 12 years to grow to 14 mm. In Massachusetts or the Wadden Sea it grows faster, reaching 26 mm in 6 years. Gilbert (1973) has proposed that growth rate and longevity are a function of the hydroclimate; with warmer temperatures a larger size and shorter life span are observed. On the Grevelingen estuary in the Netherlands, mean biomasses of *Macoma* of up to 2.93 g ash-free dry wt m^{-2} occur with a $P/\bar{B}$ ratio of 0.25 to 1.93. Low $P/\bar{B}$ ratios of 0.58 for *Macoma* have also been noted on the Forth estuary, Scotland, where a mean biomass of 2.328 g flesh dry wt m^{-2} only produced 1.365 g flesh dry wt m^{-2} yr^{-1}.

The production of *Macoma balthica* on the Ythan estuary, Scotland has been investigated in detail by Chambers and Milne (1975). They found that up to 6000 m^{-2} of new spat occurred in July, which declined in abundance to 2000 m^{-2} by December, and to about 200 m^{-2} by the following December. Animals lived for over 6 years in the Ythan, by which time they had reached over 14 mm in length. The mean annual biomass of all age classes was 4.86 g dry flesh wt m^{-2}, which produced a total of 10.07 g dry flesh wt m^{-2} yr^{-1} ($P/\bar{B} = 2.07$).

The Minas Basin, Bay of Fundy, Canada has tides of up to 17 m range. On the intertidal mudflats exposed by the tide, *Macoma balthica* is the dominant organism, along with the amphipod *Corophium volutator*, and adult *Macoma* can be as abundant as 3500 m^{-2} which is the highest density recorded from North America. In the study at Minas Basin, the abundance of *Macoma* was compared with environmental factors such as sediment grain size, tidal elevation, organic carbon and nitrogen, and bacterial density; the sediment grain size was found to set limits for the animal, with none being found in sands coarser than 0.23 mm, or finer than 0.024 mm. However, within these limits it was found that the density of *Macoma* could be accurately predicted by bacterial density alone. *Macoma* cannot survive by deposit-feeding on bacteria alone, and to survive in the large densities reported at Minas Basin it must supplement its diet by suspension-feeding.

In Cumberland Basin and Shepody Bay of the Bay of Fundy, annual *Macoma* production is up to 3.9 g flesh dry weight m^{-2} yr^{-1}, or 50% of the total macrofaunal production.

Macoma balthica in the Wadden Sea feeds as a deposit-feeding animal for up to 90% of the time that it is feeding. When deposit-feeding it is utilising mainly benthic algae which grow on mudflats. The seasonal variation in the primary production of this benthic microalga was described in Figure 2.5. *Macoma* growth starts in the spring as soon as the production of microphytobenthos reaches 100 mg C m^{-2} day^{-1} and continues throughout the spring and early summer as primary production increases. The annual growing season starts earliest in years which follow a mild winter. As the temperature reaches 10°C spawning occurs, and growth ceases at temperature above 16°C. Thus growth is restricted to a range of water temperatures between 4 and 16°C in spring. The variation in the *Macoma* population in the Wadden Sea over an 8-year period has been shown by de Wilde (1975) to be related to the level of the food supply coupled with the time of immersion by water. Recent studies by Hummel (1985) agree that *Macoma* behaves most of the time as a deposit feeder, but has shown that it depends for its food intake for the greater part on food present in the water column. The explanation for this apparent contradiction is that, as shown in Figure 3.2, much of the total food intake is rejected as being unsuitable, and is lost as faeces (or pseudofaeces). A greater proportion of the food taken in by suspension-feeding, rather than deposit-feeding, is regarded as suitable and hence it is this food which forms the main basis of their diet.

The sandflats of the Petpeswick Inlet, Nova Scotia, support a large population of the bivalve *Mya arenaria* (Figure 3.3) with over 800 m^{-2} at recruitment, falling to under 100 m^{-2} after 2 years. This population, in contrast to other populations of *Mya*, such as at Eastern Passage, Nova Scotia, lived for only 3 years, and the mean biomass of 4.57 g flesh dry wt m^{-2} had an annual production of 11.61 g m^{-2} yr^{-1}, thus giving rise to a $P/\bar{B}$ ratio of 2.54. Other longer-living populations of *Mya* have $P/\bar{B}$ ratios of about 1.0, which is a typical value for longer-living molluscs. The high ratio of 2.54 reflects the short life-span of the Petpeswick population, a feature which has been confirmed by studies of *Mya* in Swedish estuaries, where $P/\bar{B}$ ratios of up to 13.5 have been recorded for O-group animals.

The edible cockle *Cerastoderma (= Cardium) edule* lives buried within the mud, with short incurrent and excurrent siphons protruding above the surface when they are covered by the tide. Its distribution in estuaries is generally patchy and production may vary annually dependent on the

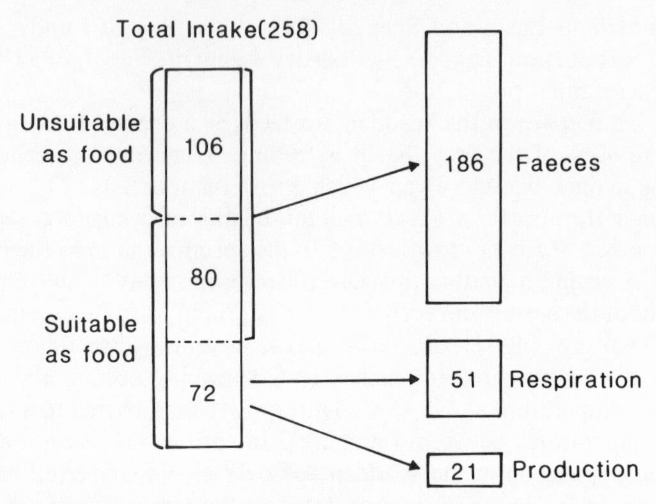

Figure 3.2 Schematic model of the main pathways in the energy budget of a population of 1 + year old *Macoma balthica* on a tidal flat in the Dutch Wadden Sea. All values are in $kJ\,m^{-2}\,yr^{-1}$. Note that of the total food intake of 258 kJ, 106 kJ were unsuitable as food, and with a further 80 kJ of food were rejected as faeces. Only 72 kJ were assimilated, for use in respiration and production. (After Hummel, 1985.)

success of the spat-fall. The newly settled spat may achieve densities of $2400\,m^{-2}$, but high rates of mortality soon reduce that number. On the Grevelingen biomasses of up to 46.7 g ash-free dry wt m^{-2} have been reported, with $P/\bar{B}$ ratios of 0.96–8.92 depending on locality. Short-lived populations, such as occur in Swedish estuaries, may have $P/\bar{B}$ ratios of 2.2–21.0. In the latter populations of *Cerastoderma*, 68% of the annual production was consumed by the shrimp *Crangon crangon* (Moller and Rosenberg, 1983).

On the Hamble Spit, Southampton Water, five species of bivalve mollusc have been recognised as key components of the ecosystems (*Cerastoderma edule, Mercenaria mercenaria, Mytilus edulis, Venerupis aurea, Venerupis decussata*). Biomass and production estimates have been made for three of them, *Cerastoderma edule* (cockle), *Mercenaria mercenaria* (quahog, hard shell-clam) and *Venerupis aurea* (golden carpet-shell). The mean biomass of the cockle was found to be particularly high with 17.8–64.6 g ash-free dry wt m^{-2} found at various stations, with an annual production of 29.2–71.4 g ash-free dry wt $m^{-2}\,yr^{-1}$ ($P/\bar{B}$ 1.1–2.6). These high values in comparison with others reported should be treated with care, as they apparently refer to selected areas within the mudflats, rather than mean

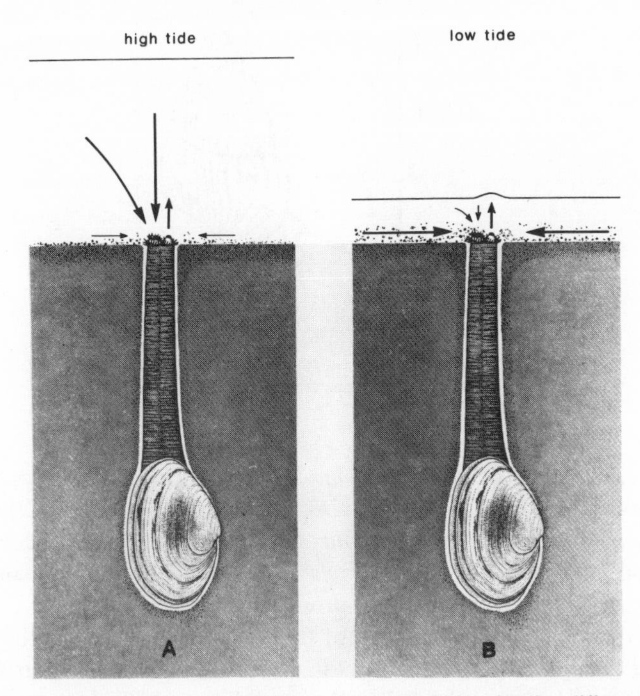

Figure 3.3 *Mya arenaria*. An adult inhabiting bottom deposits, showing different modes of feeding in relation to the height of the overlying water. At high tide the main feeding current draws in suspended material, but at low tide material is taken from the surrounding bottom surface. (From Olsen, in Rasmussen, 1973.)

values for the entire mudflat as has been case in most of the other studies discussed. *Mercenaria mercenaria* was found to have a mean biomass of 7.7–50 g ash-free wt m^{-2}, with an annual production of 4–14 g ash-free wt m^{-2} yr^{-1} (P/$\bar{B}$ 0.17–0.52), whilst *Venerupis aurea* had a mean biomass of 0.6–1.14 g ash-free dry wt m^{-2}, with an annual production of 0.7–1.25 g ash-free dry wt m^{-2} yr^{-1} (P/$\bar{B}$ ratio 1.1); however, the same caveats as those outlined above apply also to these values.

Hydrobia ulvae (Figure 3.4) is an important feeder on detritus and algae on the surface of mudflats, although it retreats into the surface layers of the mud when the tide recedes. *Hydrobia* is a selective deposit feeder which feeds mainly by grazing on diatoms growing on the surface of particles of 20–250 μm diameter. Alternatively it can secrete mucus which traps bacteria and then it reingests the enriched mucus. On the Grevelingen estuary a maximum density of 34 000 m^{-2} was recorded by Wolff and de Wolf

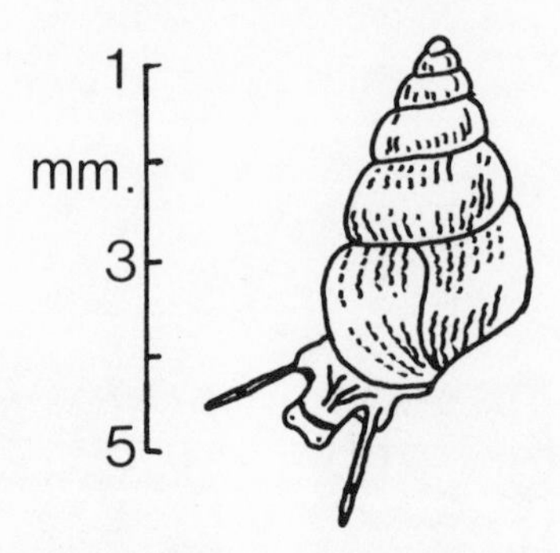

Figure 3.4 *Hydrobia ulvae*. This small snail is one of the commonest inhabitants of the surface muds of European estuaries.

(1977), with a mean density of 5147 m^{-2} representing 2.225 g ash-free dry wt m^{-2}, with a $P/\bar{B}$ ratio of 1.24–1.78. On the Skinflats area of the Forth estuary a mean biomass for *Hydrobia* of 4.7 g flesh dry wt m^{-2} occurs, with an annual production of 6.4 g m^{-2} yr^{-1}. *Hydrobia* has a life span of two years on the Forth, and the rate of production declines as they grow older. Thus in their first season (O group) the $P/\bar{B}$ ratio is 4.08, in the second season (I group) the $P/\bar{B}$ ratio is 1.90, and for the few which survive to a third season (II group) the $P/\bar{B}$ ratio has fallen to 0.0033.

Eutrophication by sewage effluent has led to the development of extensive algal mats over of the former open mudflats in several estuaries. The presence of the mat reduces the biomass and diversity of the infauna, but leads to a great increase in the numbers and biomass of *Hydrobia*. In one study at Langstone harbour the biomass of *Hydrobia* increased from 5.4 g m^{-2} to 27.4 g m^{-2}, with abundances increasing from 9000 m^{-2} to 42 000 m^{-2} due to the algal mats which provide a rich food supply for the *Hydrobia*.

The hatching of embryonic snails of *Hydrobia ulvae* can occur at salinities from 8 to 60‰, and at temperatures up to 35°C, but the optimum conditions are 37.6‰ and 19.9°C. The newly hatched snails attach to the surface of the adults shells. The wide range of possible salinity and temperature combinations for hatching coupled with the non-planktonic development and the variable modes of feeding of the adults are all important reasons for the success of this species in estuaries.

The survival strategies for estuarine molluscs are, thus, a tolerance of low salinities, a tendency towards non-pelagic larvae, and an adoption of a flexible mode of feeding designed to cope with the problems of a muddy environment. Many estuarine molluscs can utilise a number of different food sources, either at different times in their life history, or at different phases of the tide. The flexibility of feeding strategies is the hallmark of a successful estuarine mollusc.

3.2.2 *Annelids*

The polychaete annelid worms, commonly known as bristle worms, are the most diverse group of worms in an estuary. Four nereid polychaete species (*Nereis (= Hediste) diversicolor, N. (= Neanthes) virens, N. (= Neanthes) succinea, N. pelagica*) commonly penetrate into estuaries, and show differences in their physiological tolerances, and in the pattern of their life cycles. The ragworm (*Nereis diversicolor*) is able to tolerate the lowest salinities, appears to have only a brief (< 6 hours) larval planktonic phase to assist in retaining the populations within estuaries, and is usually the commonest worm in European estuaries. In America, the clamworm (*Nereis succinea*) is the most likely species of polychaete to be encountered.

The life cycle and production of *Nereis diversicolor* in the Ythan estuary, Scotland, has been investigated by Chambers and Milne (1975), who found that this population bred twice a year (June/August and January/March), producing population densities of up to 961 m^{-2}. In order to examine the size distribution of the population it was necessary to examine the size of the jaws, as the soft body of polychaete worms is liable to break or shrink under fixation. The biomass ranged from 11.49 g dry wt m^{-2} in June to 1.6 g dry wt m^{-2} in December, with a mean annual biomass of 4.22 g m^{-2} giving rise to an annual production of 12.78 g m^{-2} yr^{-1}. The $P/\bar{B}$ ratio is thus just over 3.0. In a parallel study of a brackish pond in Belgium, Heip and Herman (1979) found a $P/\bar{B}$ ratio of 2.5 in the population of *Nereis* that they studied, with a biomass of 24.3 g dry wt m^{-2} and a production of 64.2 g dry wt m^{-2} yr^{-1}, which is the highest so far recorded for *Nereis*. Forty per cent of the production of *Nereis* may be in the form of gametes, but nevertheless it is clear that the polychaetes with their shorter life spans and faster growth than, for example, most molluscs are a key component in the productivity of estuaries.

The biomass of the lugworm, *Arenicola marina* (Figure 3.5), is apparently controlled by the amount of organic matter in the sediment, although it is not found in sediments finer than a median particle diameter of 80 μm, as it cannot maintain its burrow in such fine sediments. On the Grevelingen estuary it occurs at a mean biomass of up to 8.75 g ash-free dry wt m^{-2}, with $P/\bar{B}$ ratios of 0.72–1.14. In a closely related species

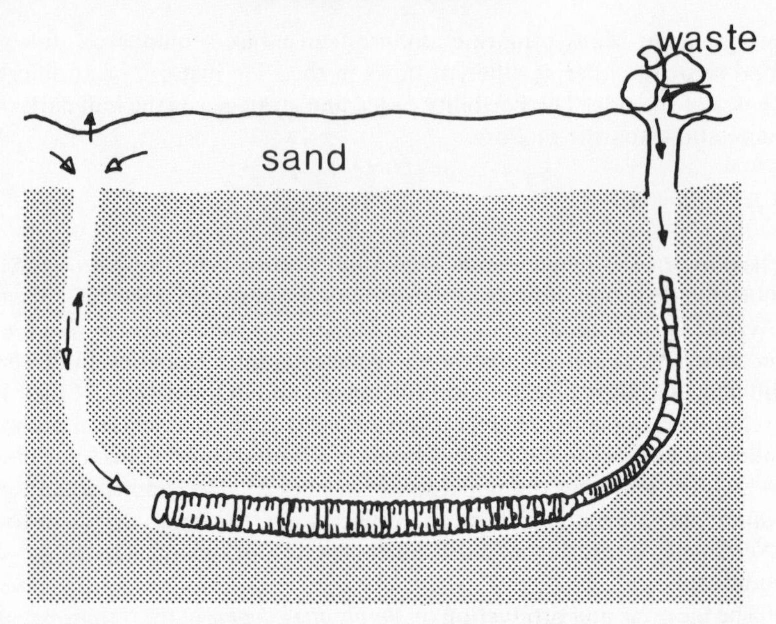

Figure 3.5 *Arenicola marina*. The position of the lugworm *Arenicola* in its burrow. Filled arrows (from waste end) indicate water currents, and open arrows (at head end) indicate the ingestion of sand. (After Kruger, 1971.)

Abarenicola pacifica it has been found by Hylleberg (1975) that a 'gardening' strategy is adopted whereby the undigested sediment which is passed out as faeces is populated by microbial organisms which increase the nitrogen content of the sediment by their activity. The enriched faeces are then reingested by the lugworm.

The sand-mason (*Lanice conchilega*) occurs on the sandier parts of many estuaries, where it forms burrows up to 30 cm deep but with a small protrusion of about 2 cm standing clear of the sand surface. For much of the time it is a surface deposit feeder, but it can spend up to 35% of its time feeding on particles in suspension in the water. This combination of two modes of feeding enables it to utilise alternative food supplies and can lead to large populations developing. In the Weser estuary, northern Germany, populations of the polychaete *Lanice* occur at 20 000 animals m^{-2} along with 700 m^{-2} of the suspension-feeding anemone *Sagartia troglodytes*. These populations have biomasses of 1090 g flesh dry wt m^{-2} and 108 g flesh dry wt m^{-2} of *Lanice* and *Sagartia* respectively. This enormous biomass of over 1.1 kg dry wt m^{-2} is possible because of the strong currents in the area (up to

$1 \, m \, s^{-1}$) carrying in diatoms and plankton for the estuarine benthos to feed upon.

The polychaete worm, *Ampharete acutifrons* is an important member of the estuarine benthos in the Lynher estuary, southern England. This is an annual species with immature animals appearing in January/February and growing rapidly until August. Heavy predation by fish, especially flounders (*Platichthys flesus*), is a vital factor in the reduction in abundance of the species in the latter half of the year. With its high growth rate the mean biomass of $0.43 \, g \, m^{-2}$ yields an annual production of $2.32 \, g \, m^{-2} \, yr^{-1}$ ($P/\bar{B}$ ratio 5.5).

The polychaete worm *Capitella capitata* is a common inhabitant of both polluted and unpolluted estuaries. In the polluted Yealm estuary, England, it has been found that the most important factor determining its distribution is the organic content of the sediment, and when this is high it can rapidly increase in abundance. It cannot digest plant material and relies on the micro-organisms within rotting vegetation for its food.

The smaller oligochaete worms have been studied much less than polychaete worms, partly due to their small size which makes them less conspicuous. However they are present in large numbers in most estuaries, thriving in the low-oxygen conditions which typify life on a mudflat. As will be seen later (Chapter 5) they may become the sole inhabitant of estuarine mudflats under conditions of organic enrichment. A study of the Forth estuary has recorded biomasses of $27.97 \, g$ dry wt m^{-2} of oligochaetes from organically enriched areas, and $6.30 \, g$ dry wt m^{-2} from more typical estuarine mudflats. The latter populations, dominated by *Tubificoides benedeni*, had an annual production of $25 \, g$ dry wt $m^{-2} \, yr^{-1}$, which was greater than the combined production of all the infaunal mollusc and polychaete populations from the same area which was $20.65 \, g$ dry wt m^{-2}. The biomass of the non-oligochaete macrofauna was $14.99 \, g$ dry wt m^{-2} which was twice that of the oligochaete, but their lower $P/\bar{B}$ ratios made the annual production less than that of the oligochaete.

The $P/\bar{B}$ ratio for estuarine annelids ranges from 1.6 for *Nereis virens*, a carnivorous species to 5.5 reported for *Ampharete acutifrons* from the Lynher estuary. A simple examination of the biomass of animals in an estuarine mudflat would tend to list the bivalve molluscs such as *Macoma* as the most important components of the ecosystem, but when consideration is given to the productivity of the fauna the annelid worms, such as *Nereis, Ampharete* and oligochaetes, with their high $P/\bar{B}$ ratios can often be seen to be more important. This importance will be reflected later (Chapter 4) in an examination of the role of worms as food for the secondary consumers.

3.2.3 *Crustacea*

The euryhaline amphipod *Corophium volutator* is an important component of many estuarine ecosystems. It lives within the upper 5 cm of the mud in a typical U-shaped burrow, emerging to collect fragments of detritus from the area around the burrow. Mossman (1978) has made a detailed evaluation of the energy flow within a *Corophium volutator* population on the edge of the Thames estuary. The population was found to consume 10 104 kJ m^{-2} yr^{-1}, of which 73.5% was rejected as faeces (7432 kJ m^{-2} yr^{-1}), 1.6% was lost as urine (166 kJ m^{-2} yr^{-1}), 14.8% was utilised in respiration (1499 kJ m^{-2} yr^{-1}) and 9.9% appeared as production (1007 kJ m^{-2} yr^{-1}). The value of production (equivalent to 95.6 g dry wt m^{-2} yr^{-1}) is very high in comparison to the biomass, and yields a high $P/\bar{B}$ ratio of 7.7 (Figure 3.6). In Swedish estuaries, 85% of the production of up to 18 g ash-free dry weight m^{-2} yr^{-1} is attributable to the first year-class of *Corophium*, and $P/\bar{B}$ ratios of between 6 and 17 have been recorded by Moller and Rosenberg (1982).

Corophium volutator is a selective deposit feeder, which can discriminate between different kinds of sand or mud. Its field distribution is the result of the choice of suitable sediments coupled with the appropriate salinity regime. It has been found that bacteria are more important than diatoms in the diet of *Corophium*, but *Corophium* can only utilise the bacteria which

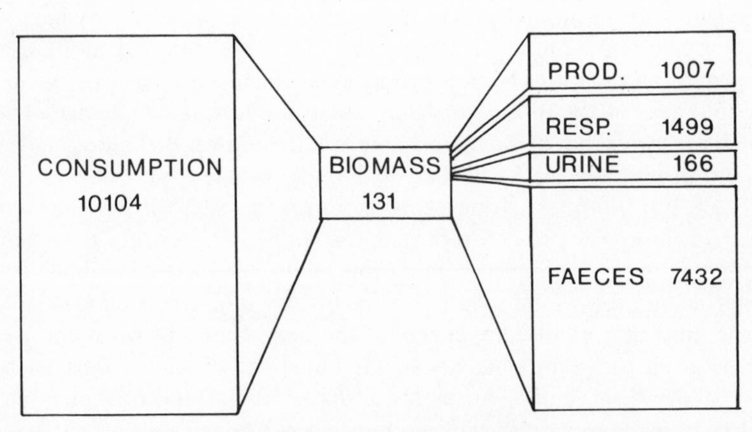

Figure 3.6 Energy-flow through the *Corophium volutator* population in Benfleet Creek, Essex, England. Units are kJ m^{-2} yr^{-1}. (After Mossman, 1978.)

are adsorbed to clay and silt particles of 4 to 63 μm diameter. Most of such particles are found in the sediment, and their presence is clearly necessary if *Corophium* is going to form its burrow within that sediment. *Corophium* is unable to filter-feed on bacteria from the water, unless the appropriate size silt particles are also present. *Corophium* often coexists in sediments with the snail *Hydrobia ulvae*, but competition is minimised by the selection of particles for feeding of different sizes, with *Hydrobia* taking the larger particles, and *Corophium* the smaller. Seafoam, which may be stranded on intertidal areas at high tide has been shown to be a valuable food source for *Corophium*, when it occurs. The seafoam traps particulate organic matter, such as algae, fungal spores, and plant fragments, all of which can be consumed by *Corophium*.

The fauna of much of the soft bottom of the Baltic Sea is dominated by the amphipod crustaceans *Pontoporeia affinis* and *Pontoporeia femorata* and the Baltic tellin *Macoma balthica*. These act as the main consumers of the detrital material which falls to the bottom of the Baltic Sea. In a study of the Baltic, at 46 m depth off Asko, where the salinity is 6–7‰, it has been found that production is dominated by the two *Pontoporeia* species, which between them produce 6.2 g dry wt m^{-2} yr^{-1} (equivalent to 339 kJ m^{-2} yr^{-1}). This contribution forms a major part of the total macrofaunal production of 6.8 g dry wt (or 150 kJ) m^{-2} yr^{-1}. This level of production is considerably lower however than the levels we have observed for the intertidal macrofauna at similar salinities in estuaries.

3.3 The surface dwellers

The edible (or blue) mussel *Mytilus edulis* develops large aggregated populations within estuaries. The accumulation of dead and live mussels typically leads to a mussel bed community developing. In the Netherlands harvesting of mussels is a major industry. From the 140 km^2 Grevelingen estuary, for example, mussel harvesters collect an average annually of 19 225 tonnes of mussels. Assuming that the meat is 20% of the total weight, and that the ash-free dry wt is 16% of the wet weight, the fisheries yield is estimated at 5.365 g ash-free dry wt m^{-2}, and the total annual production in the estuary as 16 g ash-free dry wt m^{-2} yr^{-1}. Since the mussel beds cover 14.5% of the Grevelingen estuary, this value represents an annual production of 111 g m^{-2} yr^{-1} at the mussel beds.

In the mussel bed community of the Ythan estuary, north-eastern Scotland, the mussel, *Mytilus edulis*, the gammarid amphipod, *Marinogammarus marinus* and the shore crab *Carcinus maenas* along with

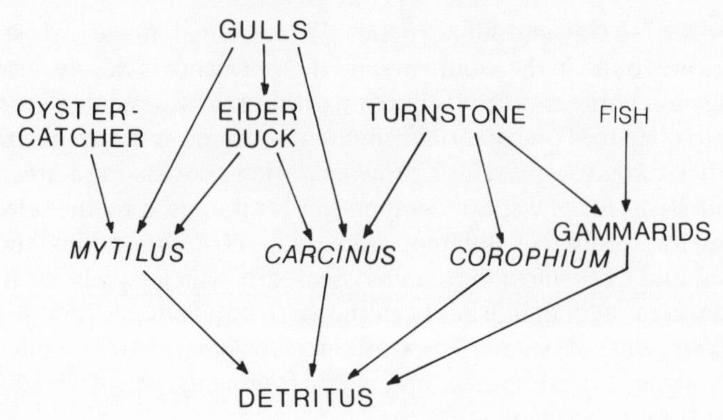

Figure 3.7 The food web of the mussel bed (*Mytilus edulis*) community of the Ythan estuary, Scotland. (After Milne and Dunnet, 1972.)

Corophium volutator are the dominant food species of the birds (oystercatcher, eider, gulls, turnstone) and the fish (butterfish, blenny) (Figure 3.7). The predators will be discussed in the next chapter, but the studies on the dominant surface-dwelling primary consumer, *Mytilus*, will be discussed here. *Mytilus* forms large beds in the lower part of the Ythan estuary, where the salinity is above 18‰, and where food is readily available, probably primarily in the form of phytoplankton and organic material of marine origin. Within the Ythan population three age-cohorts (or year-classes) can be recognised, and as discussed in Chapter 1, it is necessary to examine the production of each cohort individually. Table 3.1 shows the calculation of the production of this primary consumer population. On the Ythan the three cohorts are each preyed upon by different bird predators, with the youngest, the I group of 2–10 mm length being fed on by herring gulls (*Larus argentatus*), the two-year-old animals (II group) of mean length 18 mm, being fed on by eider ducks (*Somateria m. mollissima*), and animals three years old and over (III group) of mean length 33 mm, being fed on by oystercatcher (*Haematopus ostralegus*).

The high level of production of the *Mytilus* population of 268 g flesh dry wt m^{-2} yr^{-1} in the Ythan is of course only maintained in the mussel beds of the estuary, which occupy 18% of the intertidal area of the Ythan, and the level of production of the *Mytilus* averaged over the whole estuary is 48.7 g flesh dry wt m^{-2} yr^{-1} which is more closely comparable to other estuarine productivity studies. Growth experiments have shown that production by *Mytilus* beds may be controlled by tidal current speed through its effect on food supply, with better growth in slow current speeds than in fast speeds.

Table 3.1 The estimated gross production of *Mytilus edulis* on the Ythan estuary. Adapted from Milne and Dunnet (1972). All values in g dry flesh wt.

Age Cohort	Mean $N\,m^{-2}$	Individual mean weight increment	Production $m^{-2}\,yr^{-1}$
A Production of survivors (numbers surviving through growth period × weight increment during the growth period)			
I	2664	0.355	94.55
II	114	0.204	23.45
III	153	0.656	100.62
B Production of animals dying during period (numbers removed during growth period × half of the weight increment) (*restricted growth period)			
I	3691	0.0076*	28.33
II	207	0.102	21.19
III	no mortality detected		
Total production in study year (A + B)=			268.14

The value of 268 g flesh dry wt $m^{-2}\,yr^{-1}$ is equivalent to 5609 kJ $m^{-2}\,yr^{-1}$, assuming that *Mytilus* flesh is 20.92 kJ g^{-1}

The general reduction in current speeds within estuaries, compared to the sea, may thus benefit this animal.

Large populations of the American oyster, *Crassostrea*, are typical of the high salinity salt-marsh tidal estuarine systems common to the south-eastern United States, where they are economically important. In the North Inlet estuarine ecosystem near Georgetown, South Carolina, the oysters form extensive bars in places completely lining the intertidal creek banks. Dame (1976) has examined the energy flow in these populations where the number of oysters are between 4400 and 1000 m^{-2}. He calculated the energy assimilated by the population as 9788 kcal $m^{-2}\,yr^{-1}$, of which 4132 kcal was attributable to production (growth and reproduction) and 5656 kcal was attributable to respiration. The biomass was 2501 kcal m^{-2}, which is equivalent to 404 g flesh dry wt m^{-2}. These values for biomass and production are high, but it must be pointed out that the oyster population occupies only 5% of the total area of this ecosystem and the mean biomass for the entire area is 20.2 g flesh dry wt m^{-2}.

Oysters may have a valuable role to play in increasing the productivity of the area in which they live by filtering out food from the water column above them and depositing material as faeces and pseudofaeces which is available to the benthic organisms. The Pacific oyster, *Crassostrea gigas*, can deposit material on the bottom, equal to that resulting from the combined activity of plankton and gravitational sedimentation in a 45 m column of water, and furthermore the biodeposited particles are smaller and have a higher nutritive potential to browsing organisms. 1 m^2 of oyster bed can filter 282 720 l yr^{-1}, and retain about 11% of the food material in the water, amounting to 2570 kcal m^{-2} yr^{-1}. Much of the energy from the food filtered out from the water is utilised in metabolism and gamete production, and small amounts of energy contribute to shell growth, but the bulk of the energy ingested, amounting to 1545 kcal m^{-2} yr^{-1} is rejected by the gills or passes through the gut undigested. This material is deposited on the bottom around the oysters where it can support a large number of small deposit feeders.

The winkle *Littorina littorea* is a grazer on algal films, using its radula to remove this film from rocky outcrops or off the surface of mud and sand deposits. Within estuaries, *Littorina littorea* is very patchily distributed, and may often be found in salt marshes, eel-grass meadows and mussel beds.

In *Spartina* salt-marsh ecosystems, large quantities of plant fragments are potentially available to primary consumer animals. However, studies of the mud-snails in the salt marshes of Georgia have shown that little of the *Spartina* detritus is utilised or available as an energy source for the snails; instead they rely on the benthic algae or the bacteria which live on the plant fragments.

The production biology of the mollusc populations within the macrophyte beds of Petpeswick Inlet, Nova Scotia was studied by Burke and Mann (1974) who found that *Mytilus* was the dominant inhabitant of *Zostera* (eel-grass) beds, but it is not clear from their study how much of the food of *Mytilus* was derived from phytoplankton carried into the *Zostera* beds or how much was particulate organic fragments derived from the *Zostera*. The gastropod molluscs living within the *Zostera* were *Littorina saxatilis, Nassarius obsoletus* and *Lacuna vincta* which, it must be presumed, were grazing on plant material. Within the *Spartina* (salt-marsh) areas the numerically dominant animal was the grazer *Littorina saxatilis*. The biomass and production of these animals within the macrophyte beds are summarised in Table 3.2. The level of production of molluscs found within the *Zostera* beds are similar to those reported from benthic communities within estuarine mudflats, but the level of production of molluscs reported from the Petpeswick *Spartina* beds in rather lower than the levels of production reported from other estuarine ecosystems. It is likely that much of the production of the *Spartina* grass is carried out of the

Table 3.2 Density, biomass and production of mollusc species within the macrophyte beds of the Petpeswick Inlet, Nova Scotia. Adapted from Burke and Mann (1974).

Habitat	*Species*	*Density* $n\,m^{-2}$	*Biomass g flesh dry wt* m^{-2}	*Production* $g\,m^{-2}\,yr^{-1}$	*Production* $kcal\,m^{-2}\,yr^{-1}$
Spartina	*L. saxatilis*	331	0.81	1.5	6.0
	M. edulis	125	2.69	3.5	14
	M. lineatus	129	0.77	1.1	4.4
	Total		4.27	6.1	24.4
Zostera	*M. edulis*	817	15.2	19.7	78
	L. saxatilis	140	0.1	0.23	0.72
	N. obsoletus	35	0.89	1.15	4.6
	L. vincta	10	0.03	0.06	0.24
	Total		16.22	21.14	83.6

vicinity of the beds by tidal currents, and is utilised elsewhere in the estuarine ecosystem.

3.4 Meiofauna

The meiofauna are those animals which are able to pass through a 0.5 mm (500 μm) sieve, but are retained by finer sieves (usually 62 μm). They are thus defined mainly by their body size, and include chiefly animals with small elongated bodies which live interstitially in sand, or in the loose upper layers of mud. The most abundant of the meiofauna are the nematodes, and along with them may be found many groups of animals, especially harpacticoid copepods, and also Turbellaria, Gastrotricha, Tardigrada, Archiannelida, Coelenterata and Annelida. Earlier studies of the meiofauna tended to consider their role in the energetics of marine and estuarine benthic ecosystems as being unimportant. Latterly there has been a realisation that despite their small individual size and total biomass, the meiofauna can have a high productivity (high $P/\bar{B}$ ratio) and may contribute substantially to the production of the estuarine benthos.

In a comprehensive survey of the meiofauna of the Forth estuary, Scotland, Moore (1987) recorded a total of 172 different species, with a mean dry weight biomass for all the meiofauna of $1.1\,g\,m^{-2}$. Peak biomasses of about $4.0\,g\,m^{-2}$ were recorded close to a sewage-works and an industrial effluent. Nematodes were found to be the numerically dominant taxon, supplemented by oligochaetes in the upper estuary, and copepods and polychaetes in the middle and lower estuary. A schematic illustration of the distribution of the commoner meiofaunal species is shown in Figure 3.8. The total number of species was 50 per sample (2 × 5.5 cm^2 cores) at the mouth of the estuary, and decreased steadily within the estuary to below 10 in the upper estuary, increasing to over 15 at the freshwater head (Figure 3.9). The dip in species number in the Grangemouth area is apparently related to industrial discharges at that locality.

The most abundant representative of the meiofauna in estuaries are the free-living nematode worms. In the Exe estuary, England, Warwick (1971) has been able to recognise six major habitats for nematodes which are characterised by particular sediments and interstitial salinity regimes. Within each habitat he has listed six characteristic assemblages of nematode species. Nematodes can be divided into four feeding types, namely, selective deposit feeders, non-selective deposit feeders, epigrowth feeders, and capable of predation but probably omnivores. In pure mud and muddy-sand habitats the non-selective deposit feeders are dominant. In sand with low interstitial salinity the carnivores are commonest, reflecting the lack of suitable food for deposit feeders. In well-drained sand, food for nematodes is confined to films around the sand grains and the grazer (epigrowth) species predominate. In coarse sand which retains higher salinity interstitial water carnivores, epigrowth feeders and non-selective deposit feeders are found equally commonly.

Studies of nematodes in the Lynher estuary, UK have shown that the species diversity of the meiofauna is considerably higher than that of the macrofauna, and also that the meiofauna are able to partition the total energy available to them rather evenly amongst species, in comparison to the macrofauna. Figure 3.10 shows the partitioning of annual production amongst macrofauna, and the nematode component of the meiofauna at one estuarine site. These results suggest that for 40 species of meiofauna to coexist in 1 ml of sediment, the sedimentary environment must be highly structured on a microscopic scale, and that the organisation of meiofaunal communities must be extremely complex. The nematodes from this site have a population density of between 8 and $9 \times 10^6\,m^{-2}$ in the winter months, corresponding to a dry weight of 1.4–1.6 $g\,m^{-2}$. Summer populations reach up to $22.9 \times 10^6\,m^{-2}$ (3.4 g). Annual production is estimated to be 6.6 $g\,C\,m^{-2}\,yr^{-1}$.

An estimate of the contribution of the meiofauna to productivity of brackish ecosystems has been made by Ankar and Elmgren (1976) for sediment bottoms in the Askö-Landsort area of the Baltic. They found the

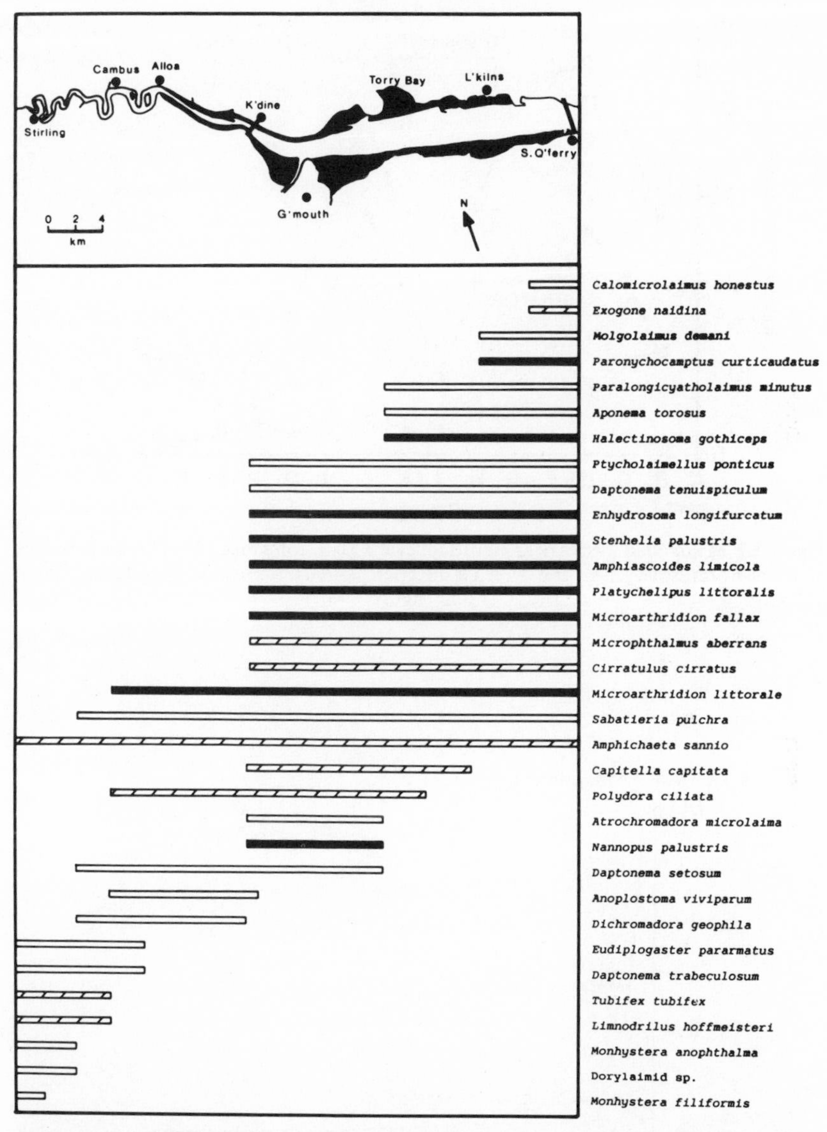

Figure 3.8 Longitudinal distribution of dominant meiofaunal species of the lower shore mudflates of the Forth estuary, Scotland. This schematic diagram shows the main area of distribution for each species, as surveyed in 1982. Open bars = Nematodes; filled bars = copepods; hatched bars = Annelids. Intertidal areas shown in black above. (From Moore, 1987.)

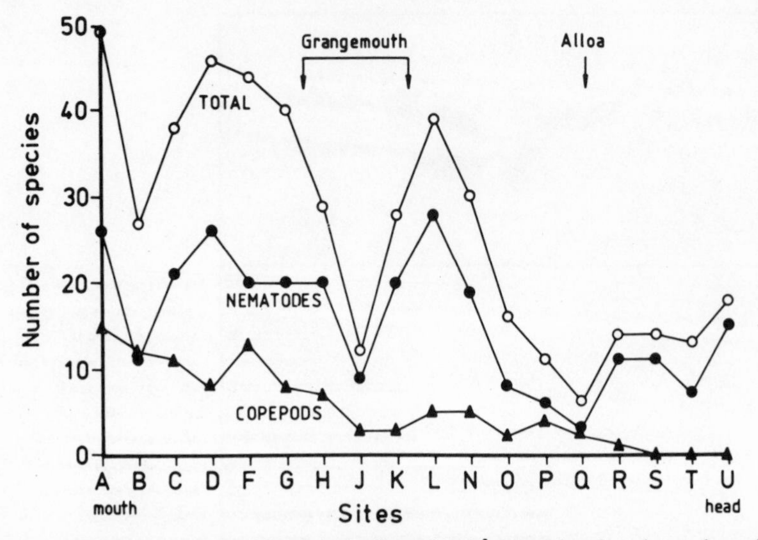

Figure 3.9 Meiofaunal species richness in duplicate 5.5 cm^2 cores taken from lower shore sites along the Forth estuary, Scotland. Note the direction plotted is the reverse of the data in Figure 3.8. (From Moore, 1987.)

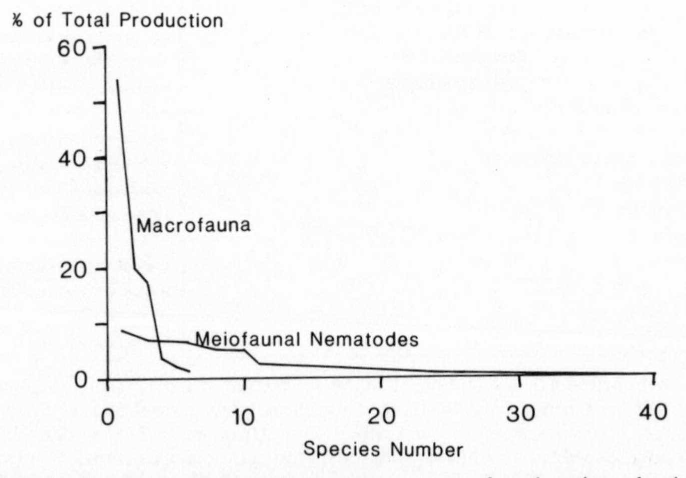

Figure 3.10 Partitioning of annual production amongst macrofaunal species and meiofaunal nematode species from a mudflat in the Lynher estuary, England. (After Warwick, 1981.)

mean meiofaunal biomass was 1.16 g dry wt m^{-2}, with the largest contribution (0.4 g) from the ostracods, whilst the nematodes contributed only 24% of the biomass. Other groups represented were Turbellaria, Kinorhyncha, Harpacticoida, and Halacaridae. Most (91%) of the meiofauna was classified as detritivorous. The production of the meiofauna was calculated as 5.5 g dry wt (or 112 kJ) $m^{-2} yr^{-1}$. This value may be compared with the production value for the macrofauna of 10.8 g shell-free dry wt (or 225 kJ) $m^{-2} yr^{-1}$ from the same area, due principally to the amphipod *Pontoporeia affinis*, along with *Mytilus edulis* and *Macoma balthica*. These values are recognised as first estimates, but they do show that the meiofauna may contribute one-third of the production within the primary consumer trophic level of a brackish-water ecosystem (Figure 3.11).

The relative importance of the different size fractions of the estuarine benthos has also been investigated on the Lynher estuary, southern England where the macrobenthos (*Mya, Nephthys, Scrobicularia, Cardium, Ampharete* and *Macoma*) had a biomass of 5.428 g C $m^{-2} yr^{-1}$ and an annual production of 5.464 g C $m^{-2} yr^{-1}$. By contrast the meiofauna (mainly the polychaete *Manayunkia*, oligochaete, nematoda and copepods)

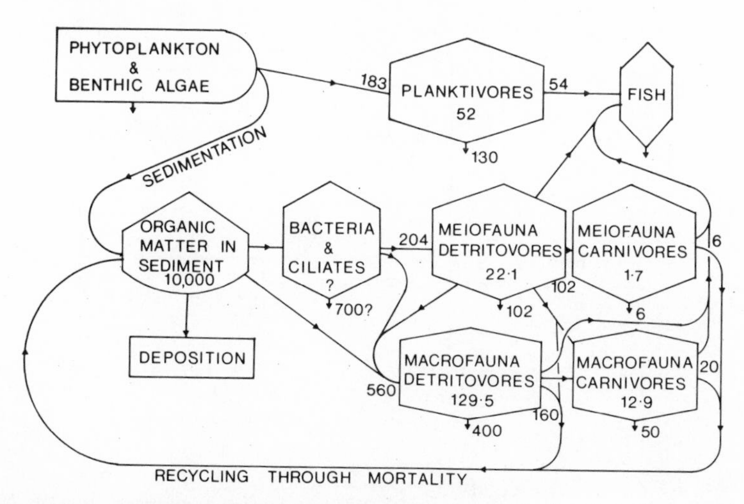

Figure 3.11 Energy-flow model (Odum energy circuit language) of the benthic ecosystem of the brackish Askö/Landsort area of the Baltic Sea. Units are kJ $m^{-2} yr^{-1}$. Figures to the left of the hexagons represent assimilation, figures within hexagons represent biomass, figures to the right of the hexagons represent production, and those below the hexagons represent respiration. (From Ankar and Elmgren, 1976.)

had a mean biomass of 2646 g C m^{-2} with an annual production of 20.391 g C m^{-2} yr^{-1}. All the macrofauna production is available to secondary consumers like birds and fish, but some of the meiofauna production is utilised by other meiofauna (especially the carnivorous hydroid *Protohydra leuckarti*); even so 16.83 g C m^{-2} yr^{-1} remains available to birds and fish. A summary of the energy flow in the Lynher estuary is given in Figure 3.12.

Bacteria, microfauna, meiofauna, temporary meiofauna and small macrofaunal elements in the benthos of estuaries and coastal areas can be regarded as one complex, characterised by the small size of the individuals, a high turnover rate (P/$\bar{B}$ ratio), relatively short life spans, and a complicated trophic structure. This grouping has been referred to by

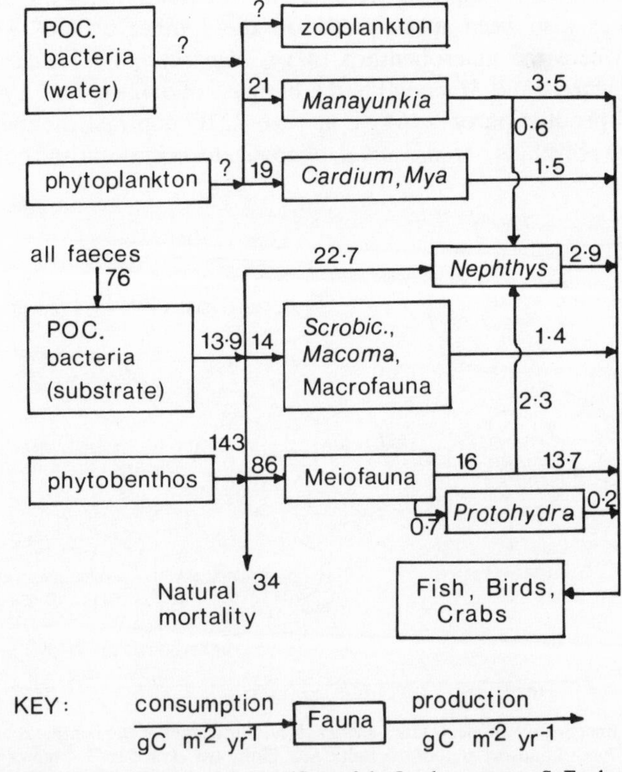

Figure 3.12 Energy flow model of the mudflats of the Lynher estuary, S. England. Units are g C m^{-2} yr^{-1}, as indicated in the key. POC = Particulate organic carbon. (After Warwick, Joint and Radford, 1979.)

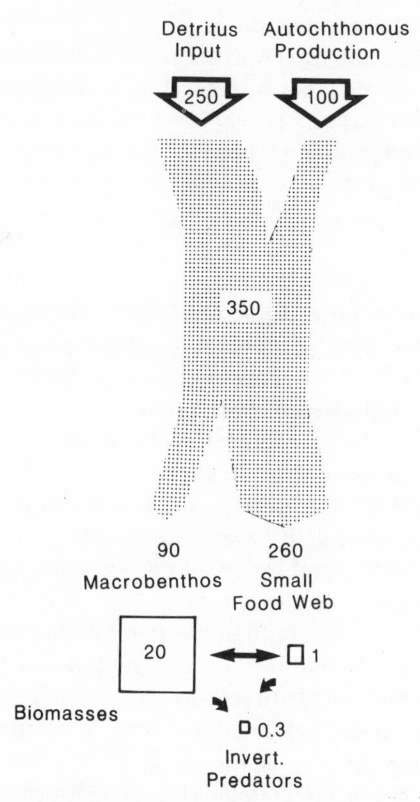

Figure 3.13 Carbon flow through the tidal flat ecosystem of the western Wadden Sea, The Netherlands. Main arrow shows the flow of detritus input and autochthonous production towards macrobenthos and the 'small food web'. Units as g C m^{-2} yr^{-1}. Boxes indicate biomasses of benthic faunal components, in units of g C m^{-2}.

Kuipers *et al.* (1981) as the 'small food web'; they have shown that in the Wadden Sea these elements may consume 70–80% of all organic material available. The carbon flow through the intertidal of the Wadden Sea is summarised in Figure 3.13, where it can be seen that of the total carbon input of 350 g C m^{-2} yr^{-1} (250 g C as detritus, 100 g C as primary production), only 90 g C m^{-2} yr^{-1} is consumed by the macrobenthos, and 260 g C m^{-2} yr^{-1} is consumed by the small food web. The mean biomasses are: macrobenthos (primary consumers) 20 g C m^{-2}; macrobenthic predators (secondary consumers) 0.3 g C m^{-2}; and the small food web 1.0 g C m^{-2}. It may thus be seen that the macrobenthos and the small food web may compete for food, and this 'biotic' factor may be the most critical limiting factor for the production of the macrobenthos, rather than any physical factor. The small food web does offer a substantial source of small food items for juvenile stages of secondary consumers such as shrimps, crabs and fishes, and may serve to explain why so many carnivores from the open sea use shallow estuarine areas as nursery areas for their youngest stages, as discussed in the next chapter.

3.5 Zooplankton

The zooplankton of estuaries is potentially limited by two features. First by turbidity, which can limit phytoplankton production and thus limit the food available for the zooplankton, and secondly (and often more important) by currents which, particularly in small estuaries or those dominated by high river flow, can carry the members of the zooplankton out to sea. The main solution to these problems is for the zooplankton to stay near to the bottom and utilise as far as possible the inflowing marine currents, but these are tidally intermittent, and the best a species may achieve is to be carried to and fro about one location. Few estimates of the biomass of the zooplankton in estuaries have been undertaken and all results are many times smaller than the rich benthic populations that we have already discussed. Within the zooplankton can be recognised two groups, the permanent members such as the calanoid copepods *Acartia* or *Eurytemora* known as the holoplankton, and the temporary members (or meroplankton) such as the larvae of benthic forms (e.g. oysters or barnacles) whose life in the zooplankton is a dispersal phase. It is well estimated that there wer 23 882 copepods m^{-3} in the Damariscotta river than the adjacent sea.

The catches of zooplankton in an estuary can be variable. The most important factor affecting variability is the tide, and therefore to get accurate estimates of the zooplankton it is necessary to take samples over a 24-hour period. Taking this into account, Lee and McAlice (1979) estimated that there were 23 882 copepods m^{-3} in the Damariscotta river estuary, Maine, with 8571 *Acartia tonsa* m^{-3}, 7360 *Eurytemora herdmanni* m^{-3} and 3753 *Acartia clausi* m^{-3}. The mean biomass of these populations of zooplankton was 0.086 g dry wt m^{-3} which with a $P/\bar{B}$ of 5, would only create an annual production of 0.43 g dry wt m^{-3}. Production of zooplankton has been measured in Chesapeake Bay, and the neighbouring Patuxent river estuary, and suggests that zooplankton production is of the order of 5–10 g C m^{-2} yr^{-1}, with the higher values in the upper estuarine reaches.

In the industrialised Milford Haven estuary, south Wales, the composition of phytoplankton and zooplankton has been investigated over a 12-year period of industrialisation. Relatively few changes occurred over the period, and the estuary continued to maintain a rich and varied plankton. Phytoplankton was found abundantly with 45 000–40 000 cells m^{-3} consisting of 42 species of diatoms and dinoflagellates. The zooplankton consisted of 6 species of medusae, 3 species of cladoceran, 18 species of copepod, 4 species of mysid, 14 species of decapod larvae and 2 species of chaetognath in 1971. The most abundant groups were larval forms both of planktonic forms (up to $6870 m^{-3}$) and of benthic animals, such as barnacle larvae at up to $2424 m^{-3}$.

In the course of a detailed study of the zooplankton of the Forth estuary, Taylor (1987) identified a total of 135 taxa, with 52 holoplanktonic taxa and 83 meroplanktonic taxa. The commonest holoplanktonic species were the calanoid copepods *Acartia longiremis, A. bifilosa inermis* and *Centropages hamatus,* followed by the cyclopod copepod *Oithona similis* and the chaetognath *Sagitta elegans.* The meroplankton was dominated by larvae of *Littorina, Carcinas maenas* and many polychaetes. Within the estuary a clear sequence could be seen with the calanoid *Eurytemora affinis* totally dominating the upper reaches, a complex of *Acartia* species in the middle and lower reaches, with *Centropages, Oithona* and *Pseudocalanus minutus* at the mouth, and penetrating further into the estuary in summer. A typical block percentage composition of copepods over the length of the estuary is shown in Figure 3.14. The trend for increasing diversity, expressed as the Shannon–Wiener information function (H'), towards the mouth of the estuary, and varying seasonally is shown in Figure 3.15.

The greatest constraint on a pelagic population in an estuary is the lack of stability, resulting from tidal mixing and high river flows. Intuitively, a true planktonic fauna would seem unlikely to develop in the upper estuary where this instability is greatest. However the copepod *Eurytemora affinis,* along with the mysid *Neomysis integer*, thrives in such situations. Both these species may be regarded as epibenthic species, they both exhibit a marked shoaling behaviour and they can apparently concentrate their distributions very close inshore at slack water. These behavioural features help them retain their position in the uppermost parts of the estuary, so that they can take advantage of a unique supply of food available there. This food supply is available at, and downstream of the freshwater/seawater interphase (FSI). The processes involved are summarised in Figure 3.16. As river phytoplankton and other material enter the head of the estuary they undergo plasmolysis, or rupture, due to osmotic stress, which liberates

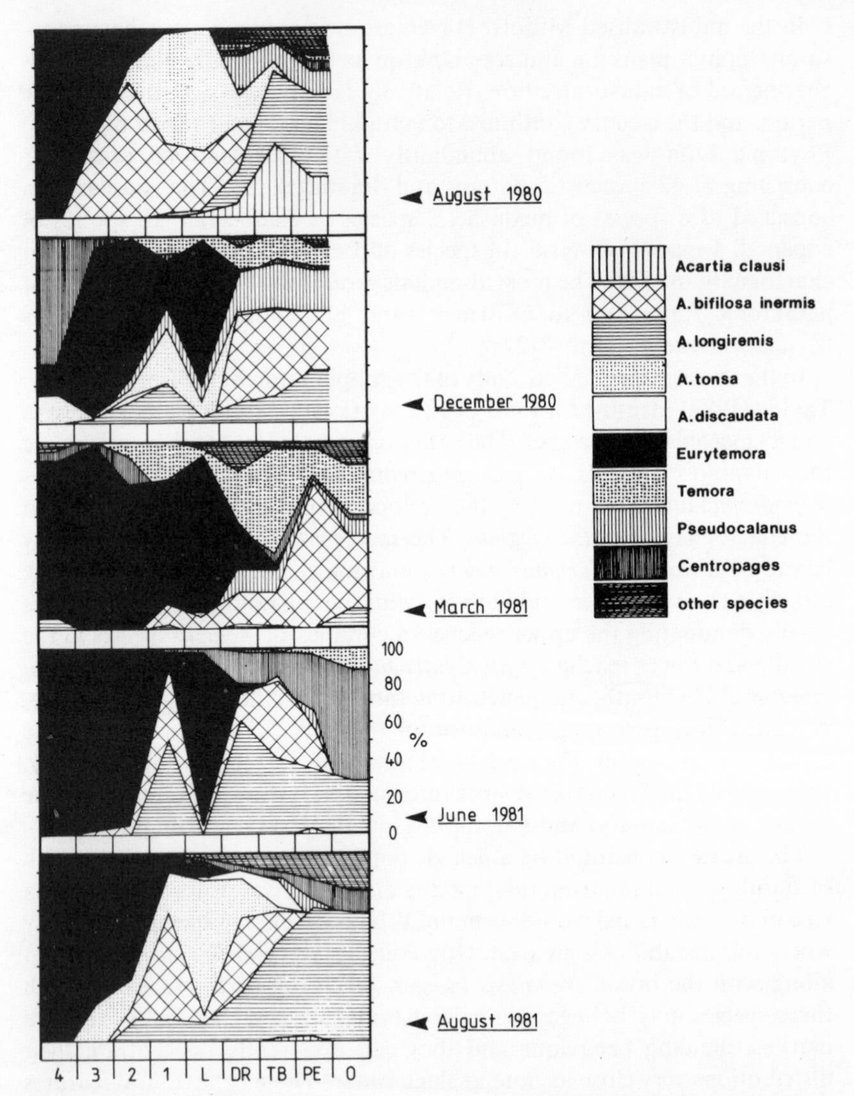

Figure 3.14 Block percentage composition of copepods over the length of the Forth estuary, Scotland at five quarterly intervals. Note that station 4 is at the head of the estuary, L at the middle, and 0 at the mouth (From Taylor, 1987.)

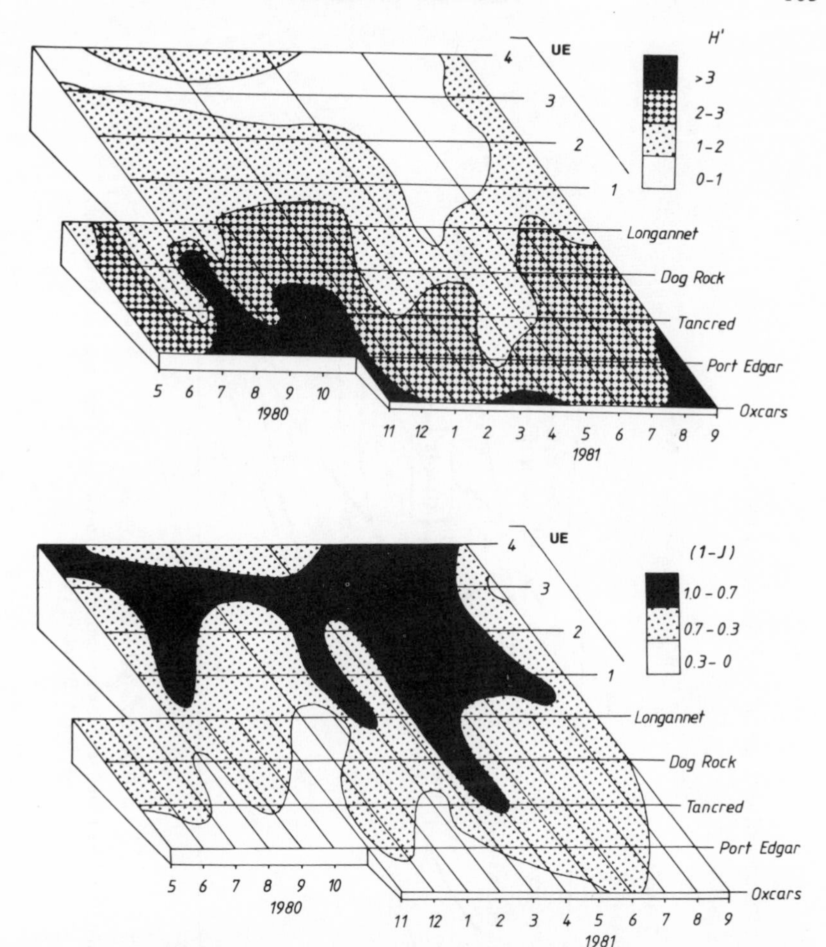

Figure 3.15 Diversity (as Shannon-Wiener index H′), and dominance (1 − J) of zooplankton species in the Forth estuary, Scotland, over the period May 1980 to September 1981. Note that station 4 is at the head of the estuary, Longannet at the middle, and Oxcars at the mouth. The diagrams indicate greater diversity, and decreasing dominance of any one species, towards the mouth of the estuary. (From Taylor, 1987.)

both soluble and particulate organic material. These organic materials are utilised by bacteria. The bacteria are consumed by ciliate protozoa, which constitute the microzooplankton, and are mainly non-tintinnid ciliates. The ciliates are then fed upon by *Eurytemora*. Addition organic matter is carried into this region by inflowing bottom marine currents, and this material experiences osmolysis in the low salinity waters of the FSI. The

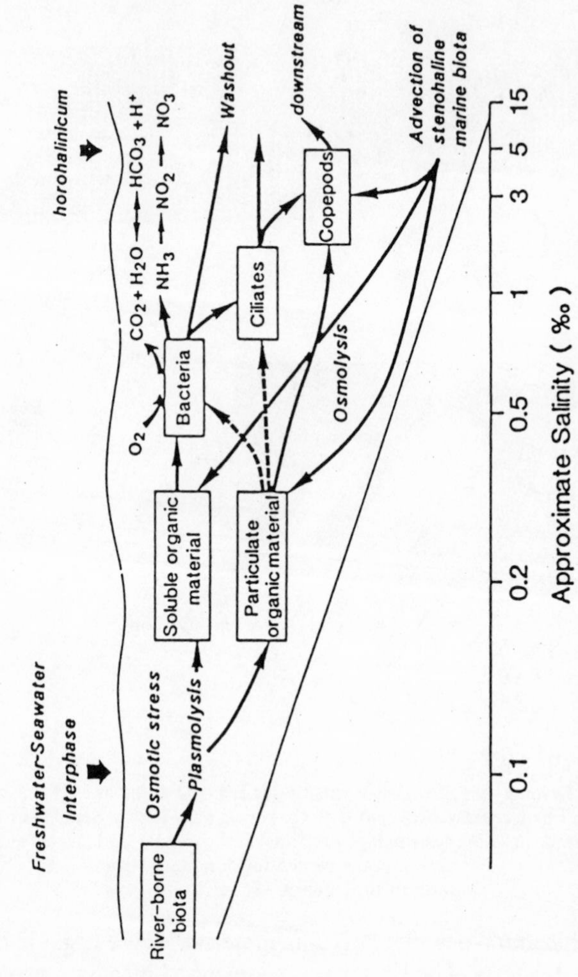

Figure 3.16 Summary of biological and related chemical mechanisms in low salinity regions of estuaries. (From P. Burkhill, PML.)

abundant growth of bacteria in this zone, utilising the organic material which is liberated here, will consume a considerable amount of oxygen, and may cause an 'oxygen sag' in these waters. The concentration of zooplankton, albeit of only one species, may thus often be greater in these uppermost reaches of an estuary, than in any other part of the estuary.

Within the Patuxent estuary, Maryland, large quantities of detritus enter the water of the estuary from the adjacent salt marshes. This occurs particularly in the months of January and February, when ice scour carries the plant debris into the estuary. Within the estuary the detritus is used by the mysid shrimp *Neomysis americana*, and also the planktonic copepod *Eurytemora affinis*, presumably in a similar manner to that described above for the FSI. The input of the detritus in the late winter has been found to generate a substantial production of *Eurytemora* of 1.3–8.65 g dry wt $m^{-2} yr^{-1}$. The *Eurytemora* in turn is consumed by fish larvae, especially those of the striped bass (*Roccus saxatilis*). This simple food chain of detritus, to *Neomysis* and *Eurytemora*, to striped bass larvae is a rapid and efficient process, and it has proved to be possible to correlate the landings of striped bass with the severity of the winters, with cold winters causing extensive ice scour and export of detritus, leading to a large population of *Eurytemora* and an increased population of striped bass.

Oyster larvae are released into the zooplankton where they selectively swim and thus contribute actively to their retention within the estuary. This swimming behaviour is connected to the increases in salinity that accompany a flood (incoming) tide, and thus helps maintain the oysters within the estuary until they find a suitable settlement site. The role of estuaries in the transport of crab larvae has been investigated in the York river estuary, USA, by Sandifer (1975) who found that the larvae of crabs which are restricted to estuaries (e.g. *Rhithropanopeus harrisii*) were adapted to be confined to the bottom waters of the estuary so that any transport that occurred would tend to be upstream and thus confine the animals to the estuary. Other more migratory crabs (e.g. *Callinectes sapidus*) produced larvae with no such adaptations so that the larvae tended to be carried out of the estuary into the sea, and later on the animals underwent a return immigration as juveniles or adults.

3.6 The primary consumer community

Although large numbers of studies have been undertaken of the biomass and productivity of individual species within estuarine ecosystems, fewer studies have been made of entire estuarine trophic levels or communities.

Table 3.3 Estimated biomass and production of four communities on the Spurn Bight mudflats of the Humber estuary as an example of the calculation of total community production. Figures are given only for the main species. (From Key, 1983, in Jones, 1988.)

	Biomass (*t*)	*Production (annual)* (*t*)
SPARTINA ZONE (91 ha)		
Hydrobia ulvae	14.1	14.1*
Hediste diversicolor	8.7	22.2
Edukemius benedii	1.3	—
Total	24.4 (268 kg ha^{-1})	36.4 (400 kg ha^{-1})
UPPER SHORE SILT (332 ha)		
Macoma balthica	59.6	123.9
Hydrobia ulvae	13.1	13.1
Hediste diversicolor	12.7	42.9
Total	91.5 (276 kg ha^{-1})	182.3 (549 kg ha^{-1})
MIDSHORE MUD (1747 ha)		
Macoma balthica	321.0	229.0
Cerastoderma edule	134.0	80.5
Nephthys spp.	64.5	147.0
Hediste diversicolor	25.3	83.4
Hydrobia ulvae	19.0	19.0
Retusa obtusa	5.1	24.0
Total	571.3 (327 kg ha^{-1})	582.9 (334 kg ha^{-1})
DOWNSHORE SAND (786 ha)		
Macoma balthica	32.0	8.3
Nephthys spp.	9.3	21.3
Cerastoderma edule	3.9	2.4
Total	46.9 (59.7 kg ha^{-1})	32.2 (41 kg ha^{-1})
TOTAL FOR WHOLE AREA (2956 ha)	734.1 (248 kg ha^{-1})	833.8 (282 kg ha^{-1})

*Productivity figures were not available for *Hydrobia*, this estimate assumes a P:$\bar{B}$ ratio of 1 and is, therefore, likely to be a considerable underestimate. All these figures are liable to various degrees of under- or over-estimation and exclude *Mya* and mysids etc. In general, the upper shore sites are likely to be underestimated whereas the lower shore figures are more likely to be overestimates.

Overall estimates are compiled by adding the biomass and annual production for the different communities or zones within the estuary. In Table 3.3, as an example of such compilations, the biomass and production of the Spurn Bight area of the Humber estuary is given. In Table 3.4 the total biomass and production of several estuaries are given.

The levels of production for estuaries tabulated in Table 3.4 are much

Table 3.4 Biomass and production values for some estuarine macrobenthic assemblages. All values as g (ash-free dry weight) m^{-2} (yr^{-1}) unless otherwise stated. Data from Wolff (1983), Knox (1986), McLusky (1987), Jones (1988).

Area	*Biomass* $g\,m^{-2}$	*Production* $g\,m^{-2}\,yr^{-1}$
Long Island Sound, USA	54.6 (dry)	21.4
Kiel Bight, Germany	26.3	17.9
Lynher Estuary, England	13.0	13.3
Southampton Water, England	90–190	152–225
Grevelingen estuary, Netherlands	20.8	50–57
Forth estuary, Scotland	10.5	12.9
Humber estuary, England	24.8	28.2
Upper Waitema Harbour, New Zealand	17.9	27.3

higher than reported for marine areas, where $4.5\,g\,m^{-2}\,yr^{-1}$ has been recorded for Loch Ewe, Scotland, and $1.7\,g\,m^{-2}\,yr^{-1}$ has been recorded for the North Sea benthos, off Northumberland. The highest values recorded from freshwater localities are the studies of the eutrophic Loch Leven, Scotland, where the bottom community produced $525\,J\,m^{-2}\,yr^{-1}$ (equivalent to 26.3 g ash-free dry wt $m^{-2}\,yr^{-1}$).

Bloom *et al.* (1972) have examined the relationship between assemblages (or communities) of animals and the nature of the sediment in the Old Tampa Bay estuary, Florida. They found an inverse relationship between deposit feeders and filter feeders, suggesting that filter feeders prefer a sediment of an optimal median size, and that deposit feeders prefer sediments with a high organic content. In analysing the distribution of the fauna and its organisation into faunal assemblages or communities, they came to the conclusion that the communities which were observed in their estuary are statistical abstractions from continua of distributions of the member organisms, rather than communities which were biologically organised, ecological entities. Thus it may be suggested that it is the factors which control the distribution of the individual species which are responsible for the assemblages that we observe in estuaries rather than some process of biological organisation. In assessing the total productivity of an estuarine ecosystem we should thus remember that whilst it is useful to group animals together in order to evaluate their contribution to the next trophic level, explanations of the reasons for the distribution of a particular species should generally be sought at the species level.

The biomass of the benthic estuarine macrofauna can be heavily dependent on environmental factors, as for example, in the Byfjord, an

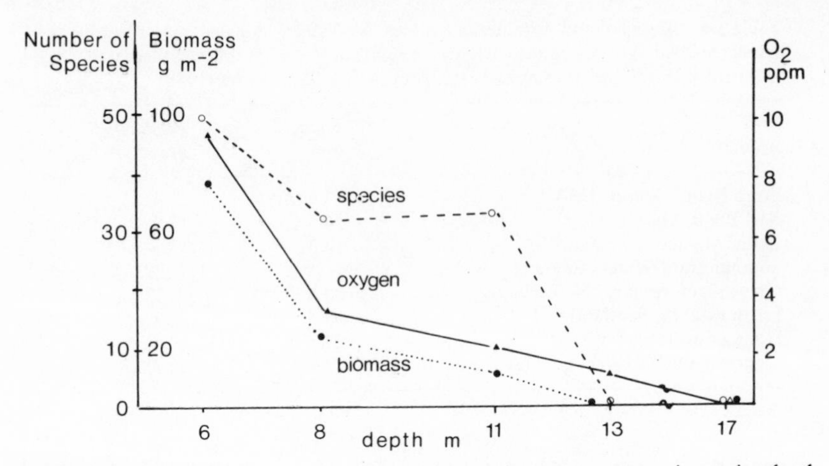

Figure 3.17 Number of species, biomass (wet weight) and oxygen content at increasing depth in the Byfjord estuary, Sweden. (After Rosenberg, 1977.)

oxygen-deficient estuary north of Gothenburg, Sweden, where both the biomass and the species diversity of the macrofauna decrease as the oxygen content decreased (Figure 3.17).

The nature of the substrate can control markedly the distribution of benthic animals on an intertidal estuarine area. Within Morecambe Bay, England, where the mouths of the Leven, Kent and Lune estuaries coalesce, the substrate above mean high water neap tide level consists of fine-grained sediments which are rich in organic carbon, organic nitrogen and phosphorus, and which are mainly inhabited by *Macoma balthica, Hydrobia ulvae* and *Corophium volutator*. Below this height the substrate is coarser with lower levels of carbon, nitrogen and phosphorus and inhabited mainly by *Tellina tenuis* and *Nephthys hombergi*. Apart from these two distinct groupings of animals, a third group was present on the beach, consisting of *Arenicola marina, Nereis diversicolor, Cerastoderma edule* and *Bathyporeia pilosa*, which were wide-ranging and occurred in variable densities. The distribution of these animals in relation to tidal height is shown in Figure 3.18 but it must be remembered that the observed distribution is caused not only by the effects of tidal exposure on the animals, but also by the tidal currents modifying the substrate which then affects the animal populations.

The tidal flats of the Dutch Wadden Sea cover an area of 1300 km^2, with a mean macrofaunal biomass estimated as 27 g ash-free dry wt m^{-2} (range 19–34 g m^{-2}). Over 90% of the total benthic macrofauna is accounted for

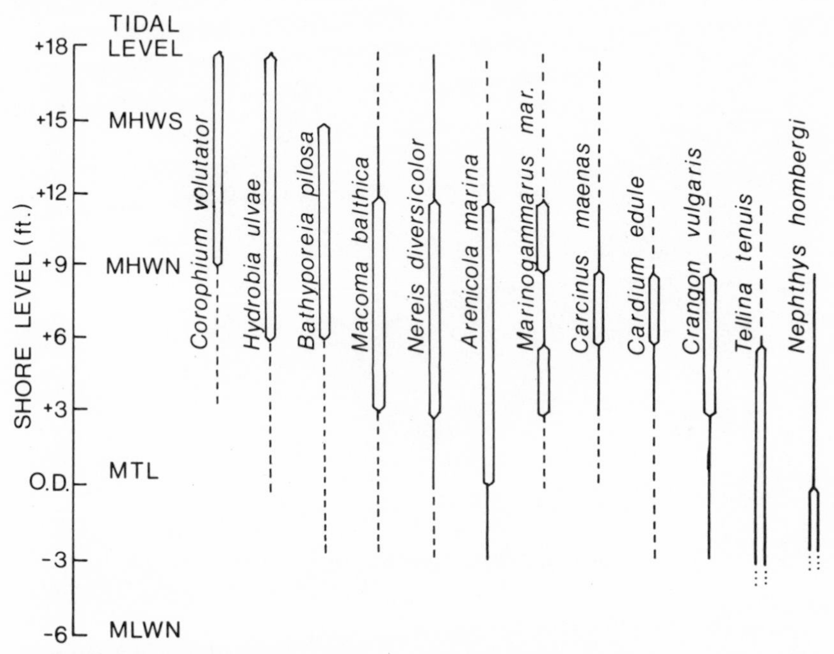

Figure 3.18 Distribution of the twelve most abundant invertebrates in Morecambe bay, England in relation to shore level. This Bay is an amalgamation of the outer reaches of five estuaries. Heights are expressed as feet (1 foot = 30 cm) above or below Ordnance Datum (O.D.). Open bars indicate densities greater than the mean, solid lines indicate densities greater than half of the mean, and dotted lines indicate presence. (After Anderson, 1972.)

by just five species (*Mytilus edulis, Arenicola marina, Mya arenaria, Ceratoderma edule* and *Macoma balthica*). Biomass has been found to increase steeply as the density of species increases. Lowest levels of biomass were found at the highest and lowest levels in the intertidal zone, and in muds and sands with the highest or the lowest level of silt content. A summary of the results of 99 transects is presented in Figure 3.19 where it can be seen that the maximum biomass (over 50 g m^{-2}) occurred between mid-tide level (MTL) and 50 cm below MTL, where the silt content was between 10 and 20%. The greatest diversity of species (over 14 per 0.45 m^2) occurred at MTL, where the silt content was between 2 and 25%. Thus it may be concluded that extreme values of environmental factors go with low biomass and low numbers of species in the mudflats of the Wadden Sea. Food was thought to be unlikely to limit the benthic fauna, and unfavourable abiotic factors were believed chiefly responsible for limiting

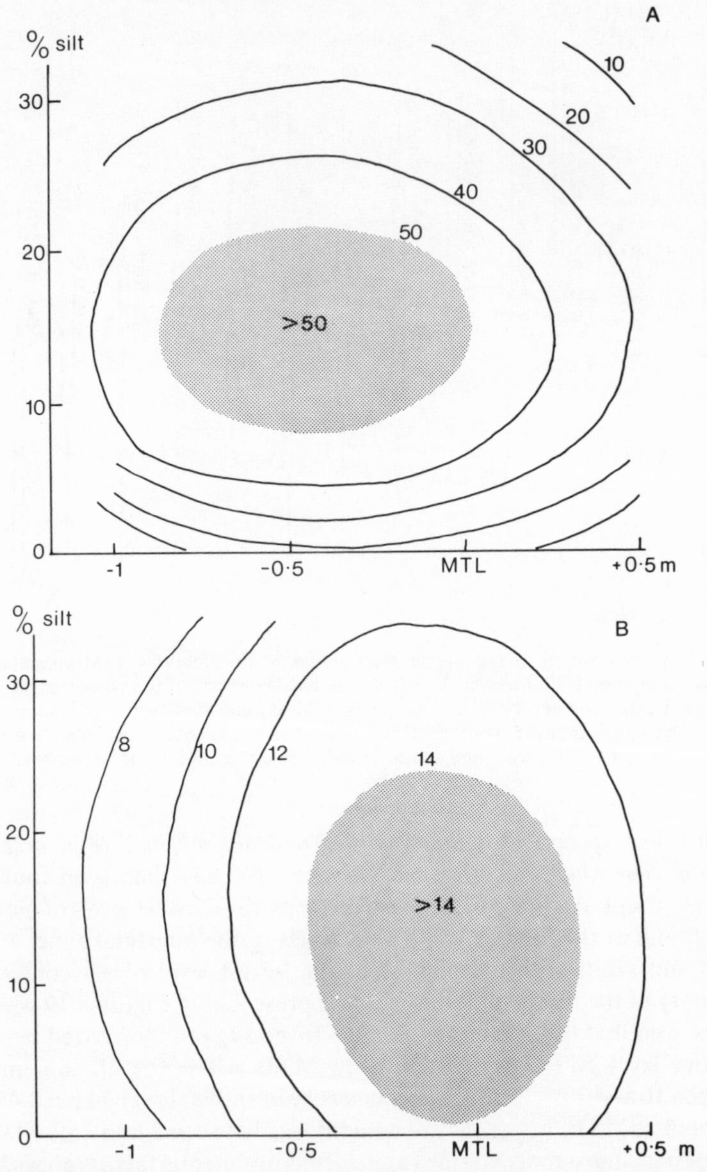

Figure 3.19 Mean values of biomass (A), expressed as g m^{-2} ash-free dry weight, and species number (B) per 0.45 m^2 on the intertidal flats of the Wadden Sea. The isopleths are drawn from the combined data of 99 transects, and expressed in relation to % silt content of the substrate and tidal height [m above and below level]. (After Beukema, 1976.)

the density of the primary consumers, however the subsequent calculations of the importance of the small food web in the same area suggest that the interception of food by these small animals, competing with the macrofaunal species may be the vital factor limiting the larger animals.

The surveys of biomass in the Wadden Sea reported above took place in 1971/2, and were repeated in 1977 showing little change in the total biomass. However the abundance of *Mya arenaria, Arenicola marina* and *Macoma balthica* had declined, whilst *Mytilus edulis, Lanice conchilega, Nephthys hombergi* and *Tellina tenuis* had all increased their biomass. These changes were related to a series of mild winters which had led to enhanced winter survival of some species, whilst poor recruitment of young had led to the observed declines. The changes reported in this study of the Wadden Sea present an excellent example of the quite large changes which take place within the primary consumer benthic community, whilst maintaining its high biomass which is characteristic of the estuarine environment.

It has been demonstrated in this chapter that estuaries can sustain high levels of production, especially among the benthic primary consumers. These high levels of benthic production are responsible for maintaining abundant secondary consumers (Chapter 4) but at this stage we are led to ask how is it that estuaries can sustain high levels of benthic production. In Chapter 2, on primary producers, we stated that two distinct types of estuaries can be recognised. In the case of American estuaries, especially those studied in Louisiana, Florida and Georgia which have small tidal amplitudes, large quantities of detritus are derived from salt marshes, mangroves or eel-grass beds within the estuary, which, supplemented by the primary production of phytobenthos and phytoplankton, produce large quantities of organic carbon which are available to the primary consumers in the estuary. Any excess organic carbon is exported by currents from the estuary to the adjacent sea. In the case of many north European estuaries, especially those studied in The Netherlands or the British Isles which have large tidal amplitudes, much of the detritus which supports the benthos is derived by importing currents from the adjacent sea or river. Seventy-five per cent of the particulate organic material in the Wadden Sea, for example, is derived from the North Sea, and only 25% is produced within the estuary by the phytoplankton or phytobenthos. The most distinctive difference is that in American estuaries studied the marsh grass *Spartina alterniflora* occupies about 2/3 of the intertidal zone, whereas in European estuaries salt-marsh plants are limited to a region above the height of high water of neap tides and 90–95% of the estuarine intertidal areas are bare mud and sandflats. Thus in the case of *Spartina*-dominated estuaries the surface-floating plant fragments will be carried seawards, whereas in mudflat-dominated estuaries the denser detrital fragments will be kept in suspension and carried into the estuary by the bottom inflowing saline currents.

Thus it can be recognised that the high level of primary production within estuaries is due to a supply of dissolved nutrients from various sources, especially by release from the sediments and by currents carrying nutrients from the sea and to a lesser extent rivers. The primary production of phytoplankton and phytobenthos is greatly supplemented by the addition of particulate organic matter either from salt marshes or seagrasses, or from adjacent seas. Due to the shallow nature of estuaries these sources of food become rapidly available to the benthic primary consumers. Sinking of surface production, coupled with tidal currents, may transport phytoplankton from the surface to the filter feeders within a matter of hours, and only a little longer time is required before the organic material sinks finally to the bottom where it becomes available to microbial decomposition and the detritus-feeding benthos.

From these considerations of the productivity of the primary consumer populations within estuaries it becomes clear that the high productivity is due to the large quantities of food material which are potentially available. Abiotic factors, such as currents, tidal exposure, sediments or salinity regimes may limit the distribution of individual species, and biotic factors, especially competition between the macrofauna and the small food web may limit the production of any one group of organisms. We have also seen in assessing the value of the primary consumers as food for the next trophic level, the secondary consumers or predators, that biomass is often not a good guide to the importance of a species. Many slow-growing bivalve mulluscs have large biomasses, but even when the heavy shell is discarded, the remaining flesh may have a low productivity as expressed by a low $P/\bar{B}$ ratio (or turnover rate). Other organisms such as annelid worms or amphipod crustaceans are faster-growing and shorter-lived and may often achieve production for the next trophic level which is over five times their biomass. The smallest organisms of all, within the small food web may have the least biomass, but the highest production. The importance of all the primary consumers as prey organisms will be discussed in the next chapter.

CHAPTER FOUR
THE SECONDARY CONSUMERS
CARNIVORES

4.1 Introduction

The secondary consumers of estuaries are many and varied; the most conspicuous are the large numbers of birds, especially waders, gulls and wildfowl, which are attracted to estuaries as feeding areas. The birds mostly feed on the rich intertidal populations of annelids, crustaceans and molluscs which are exposed by the tide. As the tide rises, many find the richest feeding at the water's edge. At high tide they may roost among salt marshes or else find other food in nearby fields if it is available. When the tide covers the mudflats large populations of fish, most typically in Europe the flounder (*Platichthys flesus*), move onto the intertidal areas to avail themselves of the food supplies there. Along with the fish come the invertebrate predators, such as the shore crab (*Carcinus maenas*) or shrimps (e.g. *Crangon crangon*). Living buried within the mud may be other carnivorous invertebrates, such as the cat-worm *Nephthys hombergi*. All of these are attracted to estuaries by the large and productive populations of the primary consumers, which are dependent on plant and detritus production which as we have seen are maintained by the ability of estuaries to trap nutrients and food particles. These rich food sources also make the estuaries less dependent on the seasonal fluctuations which characterise so many other temperate ecosystems. It is the productivity of the primary consumer which supports the variety of secondary consumers, many of which are temporary inhabitants of the estuary attracted by particular prey species at particular times of the year. A single prey species may be fed upon at different stages in its life by many different predators (Figure 4.1).

In this chapter we shall examine the impact of the various secondary consumers on the populations of primary consumers, and also how the secondary consumers share or compete for the food supplies between themselves. Almost all the estuarine birds, fish and crabs feed on the

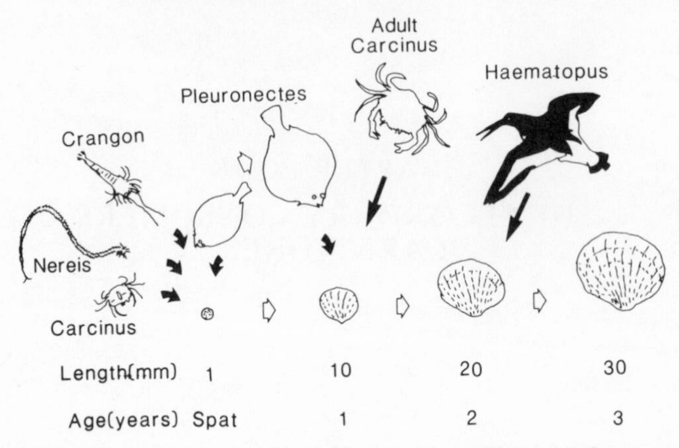

Figure 4.1 Predator spectrum of the cockle *Cerastoderma edule* growing from spat to adult size on tidal flats at the Island of Sylt, German Wadden Sea. (After Reise, 1985.)

animals of the primary consumer trophic level, but it should be noted that a few birds by-pass this trophic level and instead feed directly on the primary producer plants. Most notable of these are the widgeon (*Anas penelope*) and brent geese (*Branta bernicla*) feeding on eel-grasses and algae. Other birds such as the fish-eating birds (e.g. cormorant and merganser), or mammals such as seals, feed on the members of the secondary consumer trophic level, and should properly be called the tertiary consumers.

4.2 Fish

Fish populations in estuaries can be abundant with a wide diversity of species. Much of the abundance is seasonal as marine fish move into the estuary to breed, and having used the estuary as a nursery the young fish grow and move out to sea again (Figure 4.2). Other fish such as salmon (*Salmo salar*) and eels (*Anguilla anguilla*) use the estuary as a migratory route to get from rivers to the sea and vice versa, rarely feeding in the estuary. Only a few species live in the estuary throughout the year. Notable amongst these residents is the killifish (*Fundulus heteroclitus*) in American *Spartina*-dominated estuaries, and various species of flounder (*Platichthys flesus, Pseudopleuronectes americanus*) in mudflat-dominated estuaries. The flounders feed on the benthic invertebrates, and as we shall see later their food consumption may often rival or exceed that of the birds.

Young fish utilise estuaries and near-shore marine areas as nursery areas

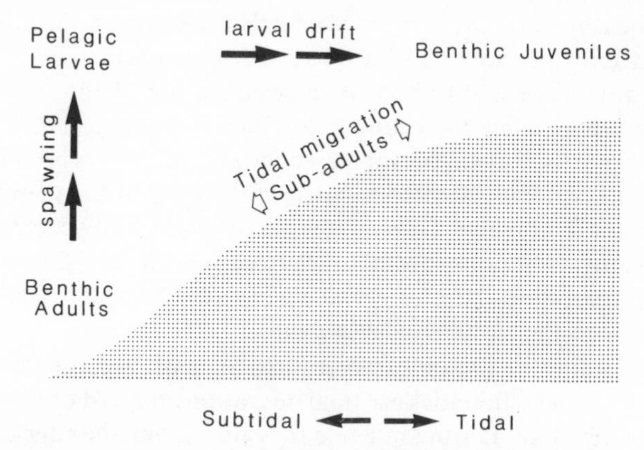

Figure 4.2 Common life-cycle in demersal fish and mobile crustaceans living in estuarine and coastal regions. The pelagic larvae develop offshore, drift inshore, and there become benthic juveniles. Later stages migrate tidally. The estuarine intertidal areas function as nursery grounds. (After Reise, 1985.)

in order to benefit from the availability of food and perhaps also to gain protection from predators. On the large estuary of the Patuxent in Maryland, for example, a clear seasonal pattern of abundance of fish has been noted with collections of fish in warmer water (spring and summer) which are abundant and diverse on the shores of the estuary, and collections in colder water (autumn and winter) less abundant and diverse and mainly in the main channel of the estuary. These differences reflect the large number of young fish which use the shore areas as nursery feeding areas in spring and summer.

Estimates of the fish population in estuaries are more difficult to make than, for example, estimates of estuarine bird populations. Some idea of the diversity and abundance of estuarine fish populations can be gained from an examination of the large numbers of fish which become trapped on the cooling-water intake screens at power stations. At the Fawley power station in Southampton Water, England, for example, fish catches of up to 60 000 per week occur. Elsewhere in Britain a massive influx of sprats (*Sprattus sprattus*) blocking the cooling-water intake screens of a power station caused damage of over £1 million.

The spawning areas of plaice (*Pleuronectes platessa*) in the southern part of the North Sea are 30–90 km off the Dutch coast. In the absence of adequate food supplies in the North Sea the very small O-group plaice swim pelagically. When they reach the Wadden Sea the presence of adequate benthic food induces the young fish to settle on the bottom. The inward movement of young plaice into the Wadden Sea, to exploit the rich food available, is brought about by a selective use of the tidal currents. Within the Wadden Sea the young plaice then consume as much as half of the invertebrate food available for all secondary consumers between March and July. In Loch Ewe, Scotland, much of the food of the young plaice is the siphons of the bivalve *Tellina tenuis*, and in any one year *Tellina* may have to regrow its siphons several times to cope with the fish predation. In the Netherlands, the tail tips of *Arenicola* form a substantial part of the fish diet, along with the siphons of *Macoma*.

Flatfish feeding in Dutch estuaries use at least two different feeding strategies. Plaice and flounder use tidal migration and feed only on the tidal flats during high tide. During low tide they wait along the edge of the tidal channel. Some individuals may swim 4 km or more per tide, but they are richly rewarded by the rich supply of prey items on the intertidal areas. Dab and sole, by contrast, are characterised by the absence of tidal migration, and remain continuously in subtidal areas feeding there. Table 4.1 compares the abundance of the major macrobenthic species at different levels of the Oosterschelde estuary intertidal areas, with the stomach contents of the plaice and flounder. In general the differences between the fish species match the differences in the faunal distribution with plaice

Table 4.1 Occurrence of seven macrobenthic species at three tidal levels on the tidal flats of the Oosterschelde estuary, compared to the average number of prey specimens in the stomach and guts of plaice and flounder. (After Wolff *et al.*, 1981.)

	Density of prey species ($\bar{x}\, m^{-2}$)			*Average number of prey* ($\bar{x}$ *per fish*)	
	High flats	*Middle flats*	*Low flats*	*Flounder*	*Plaice*
Anaitides maculata	0.0	0.0	6.7	0.0	3.1
Nephthys hombergii	0.0	11.1	4.4	0.1	0.8
Pygospio elegans	100.4	82.2	20.0	2.1	0.1
Arenicola marina—small	22.2	0.0	0.0	0.2 (small and large combined)	0.2 (small and large combined)
Arenicola marina—large	26.7	6.8	55.6		
Arenicola marina—tails	—	—	—	12.1	32.8
Lanice conchilega	0.0	0.0	4.4	0.1	0.8
Hydrobia ulvae	4575.6	7055.6	2.2	16.7	2.4
Retusa obtusa	8.9	0.0	2.2	0.0	3.4

feeding on the low flats, and flounder on the high and middle flats. In the Oosterschelde the plaice and flounder consume up to 15 g ash-free dry weight of benthos m^{-2} yr^{-1}, which may be compared to the benthic production of about 50 g ash-free dry wt m^{-2} yr^{-1}. This figure is lower than some estimates of predation by fish in other estuaries.

Fifty-four species of fish have been recorded from the small Slocum river estuary, Massachusetts over a 2-year period, with five species dominant—the mummichog or killifish, *Fundulus heteroclitus*, Atlantic silverside, *Menidia menidia*, four-spine stickleback, *Apeltes quadracus*, striped killifish, *Fundulus majalis* and sheepsheaf minnow, *Cyprinodon variegatus*. These five species dominated the estuary throughout the year, with the main summer additions being species which used the estuary as a nursery ground, such as Atlantic menhaden, *Brevoortia tyrannus* and white perch, *Morone americana*. The main factor that was found to affect the abundance and diversity of the fish populations was temperature, with maximum diversity of fish in the warmest month, July, and the least diversity in the coldest month, February. It was also found that areas with the maximal salinity variation had the most diverse fish faunas.

The killifish or mummichog, *Fundulus heteroclitus* is one of the most abundant resident fishes within the tidal creeks of *Spartina*-dominated salt marshes where they feed principally on benthic invertebrates. Studies in the salt marshes of Delaware and Massachusetts have revealed annual productivity values of 10–16 g dry wt m^{-2} yr^{-1}. The productivity values from Delaware were calculated on the basis of the populations of *Fundulus* within a 36 m long portion of a 3 km long creek. Within the 36 m portion the mean biomass was 3259–7298 g wet weight, and the population numbered 378–814 individual fish. The Massachusetts study was based on the results derived from eight tidal creeks in which minnow traps were placed, and the fish caught were marked, released, and then a proportion caught again. Such estimates of production suffer from possible errors in the estimation of the size of mobile populations, nevertheless the consistency between the two studies does point to the high productivity of these fish populations. In non-estuarine studies of fish populations, productivity values of 0.5–4 g dry wt m^{-2} yr^{-1} are typical so it can be seen that these studies of *Fundulus* are between 4 and 20 times higher. The killifish in Massachusetts have a 3-year life span, during which time they are subject to intense predation by birds and other larger fish. In contrast to many other estuarine fish species which are seasonally migratory, the *Fundulus* remain for their whole life confined to a single marsh system, apparently relying on the tides to bring their rich food supply to them. However, their sedentary behaviour has permitted a close examination of their productivity to be made, in contrast to the many estuarine fish populations elsewhere about which little is often known.

On mudflat-dominated estuaries the most common fish are often the flounder (*Platichthys flesus*) and the sand goby (*Gobius minutus*). On the Ythan estuary, Scotland, for example, the number of flounders varies from $0.05\,m^{-2}$ in winter to $0.24\,m^{-2}$ in July, representing biomasses of from 5 to 35 g wet weight m^{-2}. The total food consumption of flounders in the Ythan has been calculated as $58.1\,kcal\,m^{-2}\,yr^{-1}$, which may be compared to the consumption of gobies at $8\,kcal\,m^{-2}\,yr^{-1}$, and the combined consumption of birds (oystercatcher, dunlin, redshank and shelduck) in the same area at $23.9\,kcal\,m^{-2}\,yr^{-1}$. Thus in this estuary at least the fish populations appear to consume three times the amount of food consumed by the birds.

The gobies (principally *Gobius minutus*) of the Ythan estuary have populations of up to $0.6\,m^{-2}$ in September, representing a biomass of 0.7 g wet weight m^{-2}. The sand gobies here were found to be abundant from July until January when they were feeding principally on *Corophium volutator*, but scarce for the rest of the year. Healey (1971) in making these observations was not able to explain the cause of these variations, beyond suggesting that other fish and fish-eating birds (e.g. cormorant) might be preying upon the gobies.

The winter flounder (*Pseudopleuronectes americanus*) in Newfoundland feed mainly during the summer, consuming princiapally polychaetes, plant material, molluscs, capelin eggs and fish remains. The winter flounder is the dominant fish in the polluted Mystic river estuary in Massachusetts, comprising 89% of the total fish biomass. The winter flounder goes through its entire cycle in the estuary feeding principally on the *Capitella* worms which are the dominant members of the benthic fauna. Only two other fish species spent the entire year in this estuary, the alewife (*Alosa pseudoharengus*) and smelt (*Osmerus mordax*). Another 18 species visited the estuary seasonally, mostly in summer. The biomass of fish for this estuary averaged 2 g wet wt m^{-2}, a low value that reflects the polluted conditions in this estuary.

In a study of the fish assemblages in estuarine waters around Sweden, Thormann (1986) has shown that the number of species is limited by the salinity, with fewer species in less saline waters, but the abundance of fish was determined by temperature, with more fish in warmer waters. The fish coexist with mobile invertebrates, such as crabs (*Carcinus maenas*) and shrimps (*Crangon, Palaemon*) which feed on the same bottom fauna. In summertime the small fish have to compete with shrimps as well as other fish species for their food, and this may serve to limit the numbers of fish in an estuary. Thormann found no clear evidence that the different fish species showed any resource partitioning for their food, that is they had not evolved different diets to minimise competition between different species. Apart from competing with the shrimps and other mobile fauna for benthic food, fish such as cod and flounder also eat these animals and in a study in one shallow bay it has been shown that the fish consumed about 7% of the production of the mobile epifauna.

4.3 Invertebrates

Invertebrate predators are principally crabs and shrimps, and carnivorous polychaetes, the former mobile with the tides, and the latter buried in the mud. Shore crabs (*Carcinus maenas*) in the Yealm estuary, England, show a seasonal migration. In the wintertime they are spread throughout the estuary, but during the summer the crabs move upstream and the lower half of the estuary becomes devoid of crabs. In November they again spread throughout the estuary. This upstream estuarine migration is apparently similar to the onshore summer migration which takes place in marine crabs. In the Isefjord, Denmark, a negative estuary, the onshore migration of the shore crab (*Carcinus maenas*) has been documented by Rasmussen (1973). The crabs here move inshore in May as the temperature rises above 9°C, with the largest ones first, followed by the smaller individuals. They feed actively inshore during the summer and autumn and move offshore in October–November and hide more or less inactively in deeper water during the winter. Such migrations seem to be a clear adaptation to make use of the rich benthic production available during the summer months (Figure 4.3).

Like other estuarine animals, crabs show a sequence of different species in different stretches of the estuary. In the Avon-Heathcote estuary, New Zealand, Jones (1976) has shown that low salinity and particle size of the substrate limit the distribution of several crab species, with *Hemigrapsus crenulatus* most able to penetrate into the estuary.

Caging experiments designed to exclude crabs and gobies, and monitor the effects of these species on benthic infauna, have shown that predation by the shore crab *Carcinus maenas* can significantly reduce the abundance of small annelids on an estuarine mudflat, particularly the dominant polychaete *Manayunkia aestuarina*, and could be responsible for year-to-year variations in abundance of this species. In this experiment, on the Lynher estuary, the fish in contrast had little direct effect on the abundance of prey.

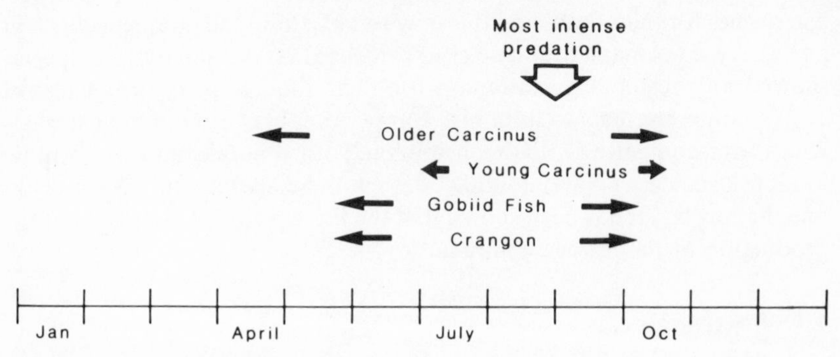

Figure 4.3 Presence of *Carcinus maenas* (shore crab), *Crangon crangon* (shrimp) and *Pomatoschistus microps* (gobiid fish) on mudflats at the Island of Sylt, German Wadden Sea. (After Reise, 1985.)

In the Lynher estuary, Devon, the carnivorous polychaete, *Nephthys hombergi* occurs abundantly with a mean annual biomass of 3.947 g m^{-2} and an annual production of 7.335 g m^{-2} yr^{-1}. Since populations of other macrofaunal invertebrates were inadequate to support this level of production, it is suggested that much of the food of *Nephthys* must be meiofaunal. The carnivorous polychaete, *Nereis virens* often forms substantial populations, feeding on smaller polychaetes. The ragworm *Nereis diversicolor* has been shown to be a predator on *Corophium volutator*, small *Macoma balthica* and chironomid larvae, as well as an omnivorous feeder on detritus.

Predation also occurs within the zooplankton, for example the planktonic sea gooseberry, the ctenophore, *Mnemiopsis leidyi*, which is a carnivore feeding on smaller members of the estuarine zooplankton, can reach large biomasses through its ability to rapidly increase in population size as food becomes available. In Narragansett Bay, USA, a biomass peak of 60 g wet wt m^{-2} of *Mnemiopsis* has been reported by Kremer (1977).

The mobile shrimps and prawns (e.g. *Crangon crangon*) and the opossum shrimps (e.g. *Praunus flexuosus* and *Neomysis integer*) are often important predators in estuaries. The high densities of shrimps may often make them one of the main commercial catches in estuaries. Like crabs, shrimps tend to migrate further into estuaries and into shallower water in summertime and retreat to deeper water in wintertime. When feeding in estuaries they tend to be omnivorous, utilising plant fragments as well as smaller planktonic animals and members of the benthos. In the Dutch Wadden Sea the juvenile *Crangon* feed on planktonic copepods, but as they grow the

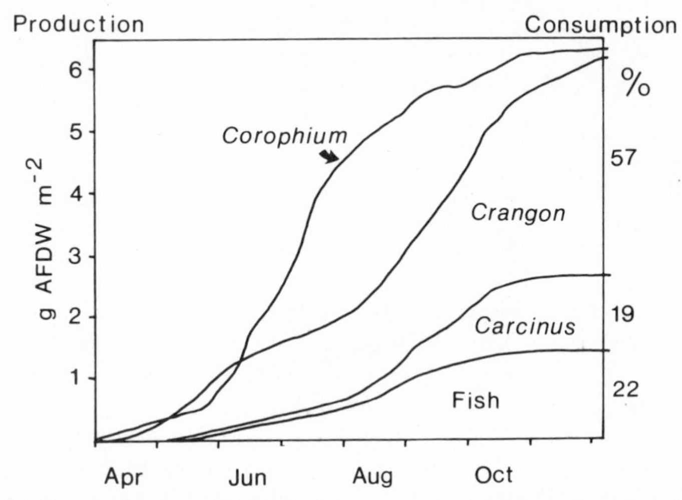

Figure 4.4 Cumulative production of *Corophium volutator*, and cumulative consumption of *Corophium* by the dominant epifaunal predators, in shallow areas of Gullmarsfjord on the Swedish west coast. Percentage consumption indicated on right. Production given as g ash-free dry weight m^{-2}. (After Pihl, 1985.)

Crangon switch to feeding on benthic foramenifera and polychaetes. In a study of the predators feeding on *Corophium* in a Swedish estuary, Pihl (1985) has shown that 98% of the annual production of *Corophium* is consumed by shrimps, crabs and fish, with 57% being consumed by *Crangon*, 19% by *Carcinus*, and the balance (21%) divided between three fish species (Figure 4.4). *Crangon* also feeds on *Nereis* and young bivalves, as well as juveniles of *Carcinus* and its own offspring. From several studies it has been concluded that *Crangon* is one of the major carnivores of shallow marine and estuarine areas, and exerts a major impact on the infaunal community.

Gizzard contents of the predatory snail *Retusa obtusa* show that the youngest and oldest feed on forameniferans, but the growing intermediate stages eat the snail *Hydrobia ulvae*. There is a close match of predator and prey size, so that young *Retusa* feed on small *Hydrobia*, and the sizes increase correspondingly throughout the summer of growth. In the Forth estuary, for example, a population of *Retusa* consumed 1.305 g ash-free dry wt m^{-2} yr^{-1} of a *Hydrobia* population which had an annual production of 5.19 g ash-free dry wt m^{-2} yr^{-1}.

The total impact of invertebrate and fish predation in estuaries is large, especially as many of the predatory species move into the estuary in summertime when the maximal biomass of prey is available. The impact of fish and invertebrates on the primary consumers will be compared with the impact of the bird predators in section 4.5.

4.4 Birds

The numbers and variety of birds of estuaries vary throughout the year. In Europe, many of the most abundant species of waders arrive in the autumn on migration from summer breeding habitats further north or east. Some of these stay in the estuaries of Britain and the Netherlands throughout the winter, whilst others move on south to overwinter in Africa. Some of the ducks such as shelduck or eider breed in the estuaries around the North Sea in summer, but they are supplemented in winter by others migrating from further north. Geese may be present in large numbers in estuaries as they pass through on migration whilst the brent geese (*Branta bernicla*) may be resident in the winter. Living throughout the year in the temperate estuaries of Europe are birds such as mallard, gulls and cormorants. In the following discussion of the role of birds as members of the secondary consumer trophic level in estuaries we shall consider their diet and methods of feeding, and in section 4.5 compare the impact of birds with that of the other secondary consumers, such as fish and crabs, on the populations of the primary consumer.

The bird predators of estuaries are highly mobile and typically exhibit clear tidal rhythms of activity associated with water movements and the activity of the prey in relation to the tides. The redshank (*Tringa totanus*) and the shelduck (*Tadorna tadorna*) which feed on *Macoma, Hydrobia* and other infaunal invertebrates feed mainly on the intertidal zone at low water, tending to be near the water's edge. Eider ducks (*Somateria m. mollissima*) feed in the shallow water at low tide, collecting in particular the mussel, *Mytilus edulis*. Others, such as the diving ducks, (goldeneye, *Bucephala clangula* or scaup, *Aythya marila*), along with cormorants (*Phalacrocorax carbo*) and mergansers (*Mergus serrator*) feed at high water by diving for their prey. In addition to the birds which feed in an estuary others such as geese, various species of gulls, certain ducks, and swans may use an estuary primarily as a place to roost and obtain their food either at sea or from land sites.

Many shore birds have adapted their food-searching behaviour to suit the tidal and seasonal rhythms of their preferred prey. It has been generally shown that the bill-length of particular species of birds is suited to particular species of prey organism, but Evans (1979) has emphasised that more subtle behavioural adaptations also occur. The bar-tailed godwit (*Limosa lapponica*) feeds principally on the lugworm *Arenicola marina*, and waits and watches for new *Arenicola* casts so that it can catch the lugworms as they defecate. The grey plover (*Pluvialis squatarola*) searches by sight for its worm prey, relying on the movement of the prey (the ragworm, *Nereis diversicolor*) at the surface of the substratum. Knot (*Calidris canutus*) on the Tees estuary feed on *Hydrobia*, and like shelduck on the Forth (see below), rely on the re-emergence of *Hydrobia* as the mudflats are re-immersed by the tide. The ragworm *Nereis diversicolor* retreats into its burrow as the tide ebbs off intertidal mudflats, but is nevertheless caught by curlew (*Numenius arquata*) relying on its sense of touch. Curlew also feed on shore crabs (*Carcinus maenas*), but as these move offshore in winter, they are only important in the summer diet.

The invertebrate prey may adjust their behaviour in response to the activity of the predators. *Macoma*, for example, in the Wash apparently remain close to the mud surface at low water whenever their predators are absent, but bury deeper into the substrate at low waters whenever predators are present. The position of burrowing animals within the substrate is most closely linked to the state of the tide, and in experimental studies of Wadden Sea animals, it has been clearly shown that *Hydrobia ulvae* and *Nereis diversicolor* migrate from being buried in the mud to the surface in direct response to the inundations of the tide over the mudflats. There is no need to invoke any inherent rhythm to the animals' behaviour, and thus the animals only come to the surface to feed just prior to inundation and when they are actually covered by the tide.

The waders are one of the most important groups of estuarine birds, especially in autumn and winter. They feed on a variety of prey organisms depending on what is available in their particular estuary. The principal prey of waders in the Wash, eastern England, is shown in Table 4.2, in which it can be seen how the prey resources have been divided between the species of waders with only limited overlap between species.

Oystercatchers (*Haematopus ostralegus*) feed on mussel beds when they are completely exposed at low tide, by hammering on one of the bivalve shells until it splits open, or inserting their bill between the valves to cut the adductor muscle. Once the shell is open, the flesh is picked out and the shell discarded. In the Wash the oystercatchers eat cockles and the daily consumption of cockles by the oystercatchers there has been estimated as 181 cockles for each bird each day (= 57.5 g total wet weight). Oystercatchers in North Wales feed on the buried bivalve *Scrobicularia plana* by

Table 4.2 Principal prey species of the main wading birds in the Wash. Data from Goss-Custard *et al.* (1977).

Bird	*Prey*
Oystercatcher (*Haematopus ostralegus*)	*Cardium edule* *Mytilus edulis*
Knot (*Calidris canutus*)	*Macoma balthica* *Cardium edule*
Dunlin (*Calidris alpina*)	*Hydrobia ulvae* *Nereis diversicolor*
Redshank (*Tringa totanus*)	*Carcinus maenas, Crangon* spp. *Hydrobia ulvae, Nereis* spp.
Bar-tailed godwit (*Limosa lapponica*)	*Lanice conchilega* *Nereis* spp., *Macoma balthica*
Turnstone (*Arenaris interpres*)	*Cardium edule* Among mussel beds
Grey plover (*Pluvialis squatarola*)	*Lanice conchilega* and various
Curlew (*Numenius arquata*)	*Carcinus maenas, Lanice conchilega, Arenicola marina*
Ringed plover (*Charadrius hiaticula*)	Not recorded
Sanderling (*Crocethia alba*)	Not recorded

thrusting their bill deep into the mud, almost up to their eyes, then jerking out the bivalve. The extracted bivalve is then wedged apart by the bill, and the soft parts removed.

Within the Morecambe Bay area, at the mouths of the Leven, Kent and Lune estuaries, northern England, is a large overwintering population of knot (*Calidris canutus*). The knot breeds in the summer in the tundra of Greenland and NE Canada, where it feeds almost entirely on insects. In winter it moves south to the estuaries and coasts of Europe and north Africa and about 16% of the total population spends the winter in Morecambe Bay. The 70 000 birds at Morecambe Bay feed predominantly on *Macoma balthica*, supplemented by *Mytilus edulis* and *Hydrobia ulvae*. The choice of *Macoma* appears to be dictated by its availability, as the knot feed predominantly on the lower half of the intertidal areas and thereby feed mainly on the *Macoma* which are generally lower down the beach than *Hydrobia* (see also Figure 3.18).

The preferred diet of redshank (*Tringa totanus*) at many estuarine sites in the British Isles is the amphipod *Corophium volutator*, and the feeding rate of redshank depends mainly on the density of that prey in mud. Where the density of *Corophium* is low, they take the polychaete worms *Nereis diversicolor* and *Nephthys hombergi* instead. The densities of both redshank and curlew (*Numenius arquata*) on the Orwell, Stour, Colne, Blackwater, Crouch and Roach estuaries in SE England are closely correlated with densities of their main prey (*Corophium volutator* and *Nereis diversicolor*). The distributions observed are due either to the birds responding behaviourally to the density of the prey in different estuaries and dispersing themselves in relation to it, or to the birds dispersing themselves in different estuaries but subsequently dying of starvation disproportionally in estuaries where food is scarce.

The relationship between bird usage of an estuary and prey density has been developed further by Bryant (1979) who has shown that for five species of wader on the Forth estuary, Scotland, there is a significant association between feeding hours (i.e. number of birds and hours spent feeding per tide) and numbers km^{-2}, and the density of at least one of their main prey. The significant associations, which explain up to 96% of the observed variation in bird distribution were found for oystercatcher with mussels (*Mytilus edulis*), for curlew, redshank and dunlin with *Nereis diversicolor*, and for knot with *Cerastoderma (= Cardium) edule*. Additionally, the distribution of bar-tailed godwit, knot and dunlin was related to the area of the intertidal mudflat, and redshank to the exposure

Table 4.3 Correlations for wader feeding-hours km^{-2} and wader feeding densities km^{-2} in relation to invertebrate prey densities and site characteristics in the Forth estuary. ○ indicates a significant association in multiple regression analysis, and × indicates items found to occur frequently in the diet in the analysis of guts. (Abbreviated from Bryant, 1979.)

	Oystercatcher	*Curlew*	*Bar-tailed godwit*	*Redshank*	*Knot*	*Dunlin*
Macoma	×○	×	×	×○	×	×
Cardium	×○	×		×	×	×
Mytilus	×○				×	
Hydrobia			×	×○	×○	×
Corophium				×		×
Nereis		×○	×	×○		×○
Nephthys			×			×
Oligochaetes				×		×
Area			○		○	○
Inaccessibility index					○	
Coverage sequence				○	○	

sequence (length of time exposed by low tide). The results of this study of bird distribution in relation to prey density are summarised in Table 4.3.

The steady fall in the numbers of dunlin, *Calidris alpina* spending the winter in Britain may be caused by the spread of *Spartina* over the mudflats where the birds feed. It has been shown that the decline in the number of dunlin varies from estuary to estuary. In estuaries where *Spartina* has not spread over the mudflats, the numbers have not changed, and the largest fall in numbers has been where the grass has spread most rapidly.

Studies on the Tees estuary, NE England suggest that up to about 90% of the macrofaunal biomass is removed by seven species of wader and duck. Most of the bird populations feed at the Tees during the winter time when the invertebrate prey is not undergoing growth, so it is therefore reasonable to compare the birds' feeding requirements with the biomass of the invertebrates. The birds select different prey items (see Table 4.4) to some extent, which enables them to live alongside each other. In the Sheepscot estuary, Maine, gulls (three *Larus* species) have been estimated to remove 6.8% of the worm *Nereis virens* population biomass, whereas in the Humber estuary, UK, only 2–3% of benthic production is consumed by birds. Clearly estimates of bird predation vary, but a value of approximately 20% of benthic production being consumed by birds can be considered as an average value.

Birds which feed intertidally in estuaries during daylight hours may find conditions particularly severe during the wintertime at high latitudes. In a study of waders in the Wash, England, it was shown that, during the autumn, feeding conditions for the birds were adequate and over 30% of the available daylight hours were spent in non-feeding activities such as resting. During the short daylight days of the winter, however, they had to spend over 95% of the available hours feeding and even that may have been insufficient to sustain many individuals who apparently died due to food shortage. During the winter the redshank on the Ythan estuary, Scotland, are only able to obtain 50% of their food from the estuary in daylight, and have to collect the balance at night or from the surrounding land.

Eider ducks (*Somateria m. mollissima*) feed by dabbling in the shallow water at the edge of the ebbing tide, or by diving into deeper water. The mussels they pick up are swallowed whole and crushed by the birds' large muscular gizzards. Such a feeding method precludes the intake of mussels larger than about 40 mm, and mussels less than 5 mm are rarely taken.

The gastropod snail, *Hydrobia ulvae*, is a major food source for wintering shelduck (*Tadorna tadorna*), and it is probable that most shelduck

Table 4.4 Importance of different invertebrates in the diet of shorebirds feeding on Seal Sands, Tees estuary, England. **Indicates the chief food providing daily biomass; *indicates regular food items;—indicates food item not taken at all or only occasionally. The total energy content of the invertebrates and the total food requirements of the birds are expressed as millions of kcals. Data from Evans *et al.*, 1979.

Shorebird	*Shelduck* Tadorna tadorna	*Grey plover* Pluvialis squatarola	*Curlew* Numenius arquata	*Bar-tailed godwit* Limosa lapponica	*Redshank* Tringa totanus	*Knot* Calidris canutus	*Dunlin* Calidris alpina	*Total energy content of invert. (million kcal)*
Invertebrate								
Hydrobia ulvae								
1 + yr class	*	**	—	—	*	**	—	9.0
0–1 yr class	*	—	—	—	—	*	*	3.0
Nereis diversicolor								
1 + yr class	—	**	**	**	*	*	*	9.4
0–1 yr class	—	*	—	—	—	—	—	4.0
Macoma balthica								
all classes	—	—	—	*	*	*	—	0.6
small polychaetes	**	—	—	—	*	—	**	85.5
small oligochaetes	**	—	—	—	*	—	**	222.0
							Total	333.5
total food requirements (millions of kcals)	73	2	23	9	17	93	68	
						Total	285.0	

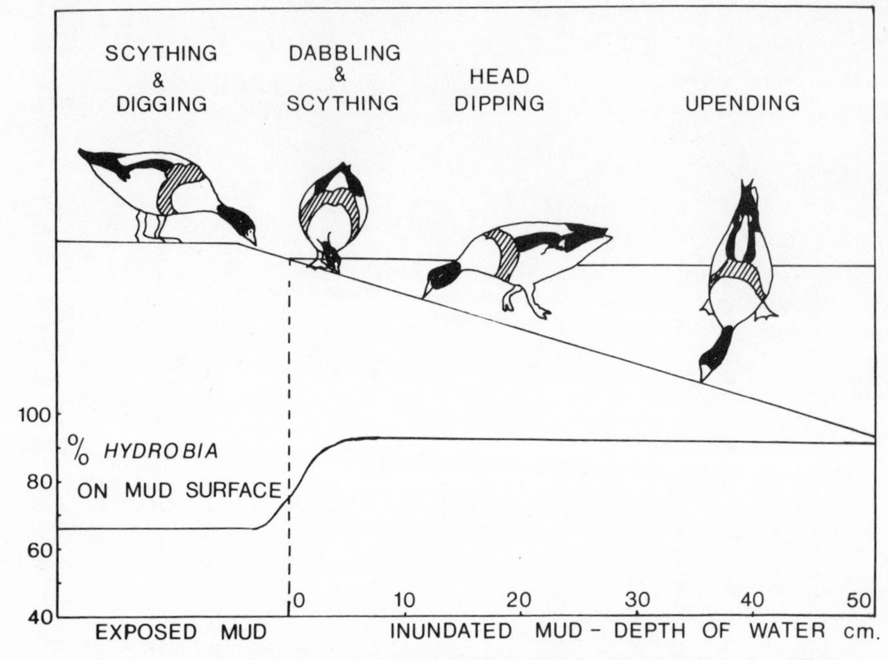

Figure 4.5 Shelduck feeding methods, showing the relationship between the different methods used and the water depth (cm), and the proportion of the *Hydrobia ulvae* population at the mud surface. (After Bryant and Leng, 1976.)

concentrations, at least in Britain, occur where *Hydrobia* is abundant and available. The feeding activity of shelduck can be classified into 5 methods, as shown in Figure 4.5. These methods are:

1. Surface digging for *Hydrobia* buried in mud.
2. Scything action on exposed mud with a moist surface.
3. Dabbling and scything in shallow water.
4. Head dipping in 10–25 cm deep water.
5. Upending in 25–40 cm deep water.

Hydrobia displays a clear tidal rhythm, with the majority of individuals burrowing into the mud surface during low tide. As the tide re-immerses the mud, the snails re-emerge onto the mud surface to commence their grazing behaviour. By adopting the head dipping and upending modes of feeding, the shelduck are able to catch the emerging snails, and to extend their time of feeding. Even at high tide many shelduck manage to continue feeding, especially by upending. The scything action of the shelduck is an effective means of sifting out *Hydrobia* and other surface-dwelling animals such as oligochaetes from the mud surface, but its value at particular locations may depend on the moistness of the mud at low tide.

4.5 The impact of the secondary consumers

The secondary consumers as a group often efficiently utilise the food which is available to them, and also partition it between themselves in response to the pressures of ecological competition. An example of this partitioning of resources has been mentioned in section 3.3, where three bird species exploit the same mussel beds of the Ythan estuary with herring gulls taking the O-group mussels, eider the I-group, and oystercatchers the II + -group. Of the summer production of 1300 kcal m^{-2} by the mussels, 112 kcal are consumed by the herring gulls, 275 kcal are consumed by the eider ducks and 93 kcal are consumed by the oystercatchers. Man also removes 240 kcal m^{-2}, and the remaining 600 kcal m^{-2} is utilised by the mussels to support their metabolism in winter. Thus it may be seen in this population that the requirements of the predators and the metabolic requirements of the mussels are all provided by the production of the mussels.

We have already seen (section 4.2) that in the mudflats of the Ythan estuary as a whole the flounders and gobiid fishes consume three times the total amount consumed by the four major bird consumers. In studies of the Wadden Sea it also appears that the fish and invertebrates are more important predators in terms of total consumption than the more conspicuous birds. In some estuaries not all of the total primary consumer production is utilised by the secondary consumers. This apparent failure to consume all the available production may be due to several reasons. Firstly, the seasonal movements of the predators due to migration in and out of the estuaries, or across continents, means that they are not always present to utilise the food available. Secondly, the limitations imposed by tidal and weather conditions may limit the time that can be spent feeding. Thirdly, much of the production of the primary consumers will be utilised after death by the decomposer organisms, and through decomposition the organic matter becomes available again to the detritus feeders.

The clear impact of predators of the intertidal macrofauna of muddy areas can be demonstrated by placing cages on the mud which exclude the predators and then observing any changes within the cages. In predator exclusion experiments in the German parts of the Wadden Sea, off the

Island of Sylt, Reise (1985) has shown a remarkable (up to 23 times) increase of abundance of macrofaunal animals within exclusion cages as compared to control sites without cages. The muddy sediments of this area are inhabited principally by the small annelid worms, *Tubificoides benedeni, Heteromastus filiformis* and *Pygospio elegans*, with the main predators in the area the shore crab *Carcinus maenas*, the shrimp *Crangon crangon*, and gobiid fish *Pomatoschistus microps*. Within cages of 1 mm mesh only a slight increase in the abundance of macrofauna was noted in the March to June period, but in the June to October period of study the numbers within the cage increased dramatically, whilst the numbers outside the cage decreased substantially. The bivalve spat within the cage grew substantially in this period, whereas extensive mortality occurred outside the cage. At the end of the study period the number of macrofauna within the cage were 86 475 m^{-2}, with the numbers outside 3750 m^{-2}. From a comparison of different mesh sizes, Reise was further able to show that in this part of the Wadden Sea the predation of shore crabs, shrimps and gobies is responsible for major changes in the intertidal macrofauna, and that the impact of birds and larger fish is negligible. The results of such cage studies should always be treated with some caution, as the presence of the cage may alter the environment.

The numbers of birds which forage on the Dutch part of the Wadden Sea are at a maximum in autumn through to early spring and then low in late spring and summer. The numbers of birds clearly fluctuates independently of food supply, as the biomass of the prey species is at a peak in July and lowest in February. The factors which control the season of peak numbers of birds must be sought elsewhere and relate to breeding migrations and the availability of food in other ecosystems. However, the numbers of fish and invertebrate predators, especially plaice (*Pleuronectes platessa*) and shore crabs (*Carcinus maenas*) in the Wadden Sea are closely linked to the maximum availability of their food species for as the biomass of prey species reaches a peak in June through to September so do the numbers of plaice and shore crabs.

In general, within the temperate mudflat-dominated estuaries of Europe there is a seasonal variation in the biomass of potential prey organisms with a maximum during the summer months. The migration of fish and invertebrate predators into the estuaries is often synchronised with this peak in the availability of the prey, and when calculated on an annual basis the fish and invertebrates are often clearly the main secondary consumers. Many estuaries particularly in Britain and The Netherlands see peak numbers of birds in the winter months, as the waders and wildfowl migrate there from eastern and northern Europe. Their areas of summer residence show an even greater seasonal variation in food availability with a rich abundance in the summer, and virtual absence of food in the winter. The birds are thus driven to the estuaries of milder regions, where even though food may not be at its maximal biomass, it is at least available. The availability of prey species throughout the year is ultimately due to the dependence of the primary consumer on detritus as their main food source since this, almost uniquely amongst primary food sources, is available throughout the year, albeit at a somewhat reduced level in the wintertime.

We have seen that the primary consumers of estuarine intertidal areas are heavily exploited by the secondary consumers which visit the area. These visitors are primarily young fish, little crabs and shrimps. The adults often stay in deeper water, while their larvae or juveniles come close inshore, feeding first on the meiofauna, and later on the macrofauna. Biomass is exported from the tidal flats in terms of nurslings which have

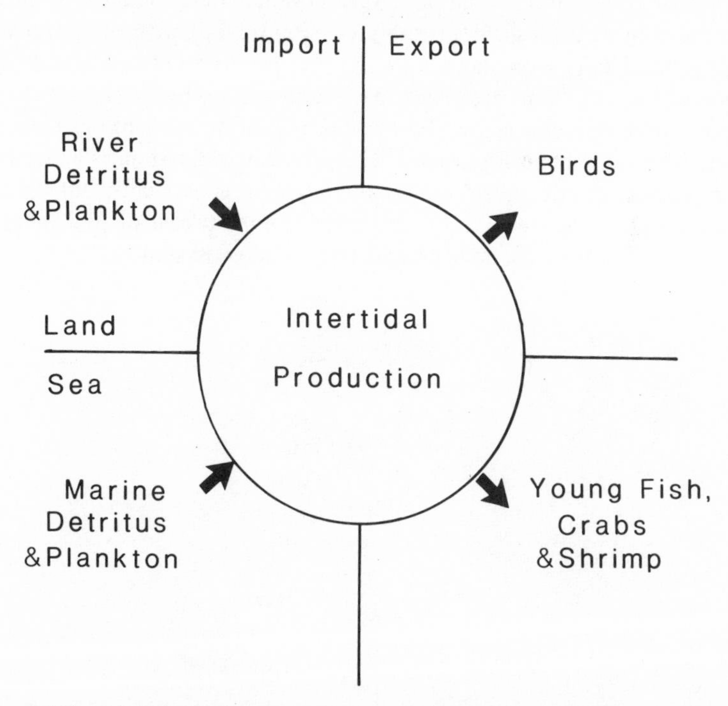

Figure 4.6 The tidal flat turntable for the flow of organic matter between the land and the sea. (After Reise, 1985.)

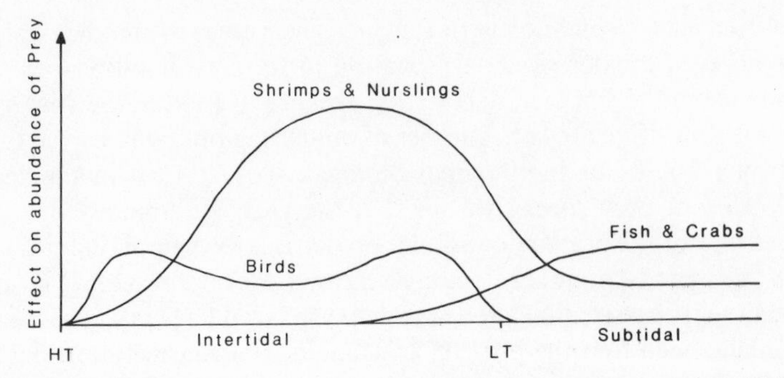

Figure 4.7 Probable distribution of predator effects on prey populations on intertidal areas from high tide (HT) to low tide (LT), and also on adjacent subtidal areas. (After Reise, 1985.)

grown to adult size and return to the sea to grow and breed, and also to a generally lesser extent by birds which feed on the same areas at low tide, and then fly off to the land at high tide. Reise (1985) in describing this situation, uses the term 'an ecological turntable' (see Figure 4.6), with energy received by intertidal flats from land and the sea as imported of detritus, and exported from the same area back to the land and sea by birds and fish. The various predator effects on the prey populations are summarised in a speculative manner in Figure 4.7. The estuarine ecosystem is as we have already seen heavily subsidised by the adjacent ecosystems, but through the activities of the secondary consumers some at least of this energy is returned to support the marine and terrestrial ecosystems.

CHAPTER FIVE
ESTUARINE POLLUTION

5.1 Introduction

The estuaries of the world receive a large proportion of the waste discharged by mankind into aquatic environments. Within the sea, almost all pollution is concentrated into estuaries and nearshore coastal zones. When environmentalists or politicians talk about pollution in the sea, they are often really discussing pollution in estuaries. What do we mean by the word 'pollution'? A useful definition is 'The introduction into the estuarine environment of a diverse range of materials, derived from human activities, in such quantities that the environment is made less suitable for existing life forms'.

As we will see throughout this chapter, the impacts of pollutants are variable. The effects of pollutants may vary according to the chemical and physical state of the material being discharged into the estuary. The effects of pollutants on the estuarine ecosystem will also vary both seasonally and temporally, as well as being related to environmental factors, such as water circulation. The effect of a particular pollutant may vary according to the part of an estuary which receives it. To take an example of such variation, a sewage effluent discharged into the head of an estuary in summer may consume all the oxygen, and bring about catastrophic changes in the ecosystem, whereas exactly the same quantity of sewage discharged into the mouth of an estuary in winter may have so little effect as to be virtually unmeasurable.

The responses of estuarine organisms to pollution range from the acute to the minimal. At the highest level of pollution, the responses of the animals and plants are easily recognised, since the results are acute and may be lethal to all forms of life. At a lower level of pollution, the sensitive fauna is eliminated, but tolerant species may thrive and become more abundant. The estuarine ecosystem thus shows distortions of conditions and organisms. Measurement of these distortions is the basis of the biological monitoring of pollutant effects. At the lowest level of pollution, only subtle changes in the physiology and biochemistry of the organisms may occur. Such subtle changes may be crucial in the long term, but can be difficult to detect.

It should be remembered that all species are tolerant of a certain amount of environmental variation, but the tolerance to pollution will certainly vary from species to species, and will often vary between individuals of the same species. It can be argued that estuarine species are already particularly tolerant to environmental variation compared to marine or freshwater species, and for this reason, the capacity of the estuarine ecosystem to accept pollutants which enhance natural variation, such as organic matter, is relatively great. The limits of environmental acceptance need to be determined, to achieve the best reconciliation between industrial development and the maintenance of amenity.

As we will see, 'hot-spots' of acute pollution are readily detectable, and often incur public wrath, but long-term, more subtle, 'chronic' pollution is often sanctioned by governments, and is less likely to feature in newspaper headlines, despite having profound effects on the estuarine ecosystem. To take an example, an oil tanker crashing into the rocks at the mouth of an estuary will command immediate attention because of the pollution risk, whereas an industrial discharge into an estuary containing only a small percentage of oil will rarely feature in the public domain, despite the fact that over the time scale of a year, for example, more oil could enter the estuary from the industrial discharge than is spilt from the stricken tanker.

In this chapter we shall first consider the uses made of estuaries, and then examine the various responses of estuarine organisms to each usage. Of necessity, we shall consider each mode of pollution separately, but it should be remembered that each impact rarely occurs alone, and a particular estuary may experience several sources of pollution.

5.2 Mankind's use of estuaries

Mankind has used estuaries in a wide variety of ways for centuries. Over one-third of the population of the United States of America lives and works close to estuaries, and seven of the ten largest cities of the world are situated adjacent to estuaries (London, New York, Tokyo, Shanghai, Buenos Aires, Osaka and Los Angeles). In Britain most of the main cities border estuaries.

Cities have developed on estuaries because of their role as natural transportation centres, providing sheltered, if sometimes shallow, harbours, and linking river and sea traffic. To supply the needs of commerce and navigation, quays and wharfs have been built along the banks of estuaries where ships can tie up to load and unload. As ships have increased in size, so the pressures on the estuary have increased. Firstly, it may be necessary to dredge the estuary to ensure a minimum depth for navigation. Secondly, as ships further increase in size, it may be necessary to extend the size of harbours or piers to accommodate larger vessels. Finally, the upper reaches of an estuary may be abandoned for navigation as the giant supertankers of the twentieth century cannot enter the old estuarial harbours, and new terminals have to be built in the lower reaches of the estuary. This process of a historical shift of estuarial harbours seawards has occurred in many of the world's estuaries, for example on the Thames estuary from London down to Tilbury and Felixstowe, on the Clyde estuary from Glasgow down to Greenock and Hunterston, and on the Chesapeake estuarine system from Washington DC down to Norfolk. The abandoned harbours may be reused to provide modern housing for city dwellers, complete with a view of their local estuary.

Along with the use of estuaries for navigation and harbours has been the use of the banks of estuaries for industrial development. Large-scale industry such as steel works, oil refineries, or chemical works need a combination of flat land and good transportation. This combination is usually found close to an estuary. Since the supply of flat land is finite, the increased demands of industry are often met by the 'reclamation' of estuarine mudflat. In extreme cases, such as the Tees estuary, UK, 90% of the former intertidal habitat has been reclaimed to provide flat land for industrial development. Elsewhere, farmers may gradually reclaim an estuarial salt marsh, by drainage, to expand their farm. Many of the harbours constructed in estuaries involve large-scale reclamation, and several 'land-fill' sites for the disposal of domestic refuse involve reclaiming estuarine intertidal areas. To prevent against flood damage whole estuaries may be shut off from the sea, as has happened in the Netherlands, for example in closing the saline Zuider Zee, to become the freshwater Ijselmeer, or in their large-scale Delta scheme.

Man has long used estuaries to dispose of waste material. Sewage is discharged into many estuaries. In many cases the raw untreated sewage is discharged, and in other cases the sewage is treated on land in septic tanks or sewage-works and only the liquid produced is discharged into the estuary. The waste so discharged may, if there is only a little, become incorporated into the estuarine ecosystem as another source of detritus.

Often, however, the quantities discharged may be so great as to cause major changes to the fauna and flora. The solid waste from sewage works, known as sewage sludge, has to be disposed of, and several countries including Britain and USA have allowed the disposal of such material to sea or estuary. In addition to sewage discharges, many of the industries situated along the banks of estuaries discharge their effluent, treated or not, to the adjacent estuary. Such industrial effluents may range from fresh water or sewage, through to a complex variety of chemical wastes.

The dredging of the upper and middle reaches of estuaries to provide shipping access to a harbour generates large quantities of spoil material which must be disposed of. Sometimes such dredge spoil will be used for infilling at a reclamation site, but often it is transported by ship and dumped in another part of the estuary. Thus not only is the dredged area affected, but also the dump site. To supply the needs of construction industries for sand and gravel, such materials are often dredged from the bed of the lower reaches of an estuary.

To generate the electricity required for a modern city requires the construction and operation of large power stations. Whether powered by coal, oil, or nuclear fuel, power stations need large volumes of water, both to heat for use as steam in turbines and also to cool the steam after use. Many such power stations have been built on the banks of estuaries, and the large volumes of water used generate large volumes of waste water. At one time, such waste water was considerably warmer than the estuarine water it was discharged into, but in modern power stations everything is done to conserve heat, and the waste water produced should be close in temperature to the water of the estuary. If coal is the fuel used, then the ash produced must be disposed of, and in several locations ash disposal has resulted in estuarine reclamation. If nuclear fuel is used, then concern may be expressed about any radioactivity in the waste water discharged. Oil is extracted from several estuarine locations, most notably the Nigerian oilfields are located within Nigeria's estuarine delta region, and the risks of oil spillage need to be guarded against.

Apart from the catalogue of ways in which man has potentially abused estuaries, mention should be made of other uses, such as for fisheries, recreation or conservation. Worldwide, the main estuarine fisheries are for shellfish, with the collection of some of the abundant natural populations of invertebrates, such as shrimps, crabs, oysters, cockles and mussels. Increasingly the catching of natural stocks is supplemented by mariculture, most notably for oysters and mussels. Bait digging by sport fishermen occurs widely in estuaries, and may involve considerable disruption of the intertidal fauna.

In addition to the shellfish industries are the vertebrate fisheries. Whilst many of the fish which enter estuaries are not commercially exploited because they are the nursery stocks, or unwanted species, other species are heavily exploited. Salmon, sea trout and eels all pass through estuaries on route from the sea to rivers, and many commercial fisheries exploit them. The Gulf of Mexico, stretching from Florida to Texas and Mexico, is a key fishery area for the United States of America, with 28% of their total fish landings coming from the brackish bays and lagoons of this area. The catch is mainly menhaden (*Brevoortia* spp.), striped mullet (*Mugil cephalus*), croaker (*Micropogon undulatus*) plus Penaeid shrimps, blue crab and oysters. All except the croaker are estuarine species, which in general spawn at sea. The larvae enter estuaries, grow there, and when fully mature return to the sea. Other rich estuarine fisheries are at the mouth of the Amazon, in Nigerian estuaries, and in Indian estuaries, such as the Ganges, where estuarine conditions extend 160 km upstream from the mouth.

Recreational interests in estuaries range widely; the sheltered waters are used for sailing, swimming or wind-surfing, often accompanied by the construction of specialist harbours or marinas. Many people are attracted to the tranquility of estuaries for bird-watching or simply escaping from their fellow men. The bird-watching fraternity have in particular become a powerful pressure group, campaigning for the conservation of the remaining parts of estuaries. In the next chapter we will discuss the ways in which estuarine conservation may be managed.

We can thus see that the uses of estuaries by mankind are many and varied. In almost every estuary a variety of uses may be occurring, and many of the uses described are not mutually exclusive. Obviously some conflicts do arise between the different users of the estuary. The requirements of fishermen or conservationists for clean undisturbed waters may conflict with an industry or city council looking for the cheapest disposal site for waste material. Despite the many uses of an estuary, for most people living in cities their local estuary is the closest natural ecosystem that they can visit or observe, since the resilience of estuarine fauna has meant that it has survived, whereas other natural habitats, such as the countryside in all developed countries, have been irrevocably altered.

5.3 Methodologies of studying pollution

All organisms respond to environmental change from whatever cause by a series of responses ranging from subtle metabolic adjustments to dramatic changes such as escape or death. Blackstock (1984) has usefully brought together the time-related sequence of the effects of environmental stress on

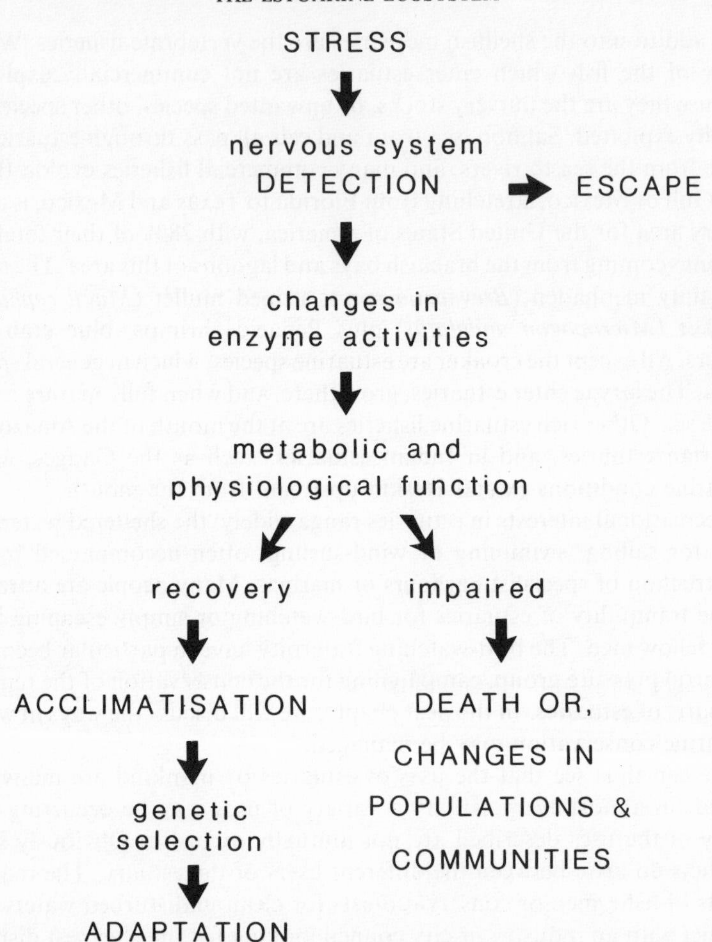

Figure 5.1 Schematic sequence of the effects of environmental stress on estuarine animals. (After Blackstock, 1984.)

marine and estuarine fauna in a single figure (Figure 5.1). The first event of importance to the organism is the detection of change by the sensory receptors, followed by metabolic adjustments and/or behavioural reactions. Mobile animals can swim away from the affected area, animals within the sediments may burrow deeper, and stationary animals such as mussels may simply shut the valves to wait for improved conditions.

Depending on the animal and the cause of the stress, the animal may recover its metabolic functions, showing acclimatisation, or individual animals may be genetically selected to survive the stress, showing adaptation. Alternatively the stress may cause death or serious impairment of normal functions, such as growth or reproduction, leading to changes in populations and communities. Such structural and functional responses can be detected by biological sampling of the components of the estuarine ecosystem.

A wide variety of techniques has been developed to study the changes which occur in estuarine organisms in response to pollution, ranging from laboratory-based toxicity testing procedures, through to field-based population sampling programmes, and including physiological, cytological, biochemical, genetic, behavioural, pathological methods as well as bioassays and ecologically-based methods. Bayne *et al.* (1985a, b) have described many physiological, biochemical and cytological techniques which have been used to measure the biological effects of environmental stress factors, including principally pollution. Davenport (1982) and Lockwood *et al.* (1982) have described techniques to simulate the fluctuating environmental conditions of an estuary, whilst simultaneously exposing the animals to pollutants. So for, however, such techniques have not been widely used for routine pollution assessment.

Toxicity testing of a pollutant involves placing the animals in static, or preferably flow-through tanks, with known concentrations of the pollutant. The results can be expressed in different ways, but all follow the basic concept shown in Figure 5.2, where response is related to dose. For a non-essential element, small doses produce little response, but as the dose increases the response increases. Below a certain threshold the responses are considered as acceptable changes, but above that level the responses are considered unacceptable. For an essential element small doses are vital for life, and lack of the small dose, as well as large doses, produce unacceptable change. In between these dosages is an ideal dose. All toxicity testing is designed to determine the safe threshold between acceptable and unacceptable change.

In acute toxicity testing the median lethal time, LT_{50}, that is the time at which half the population die at a given dosage is measured. From the LT_{50} can be derived the median lethal concentration, LC_{50}, for a specified time, conventionally 96 hours. Since such tests only determine the lethal (or acute) stress of a pollutant, it is common to multiply the results by an application factor (0.1 or 0.01) to determine a 'safe' level for the pollutant. Lethal tests are widely criticised, and many investigators recommend

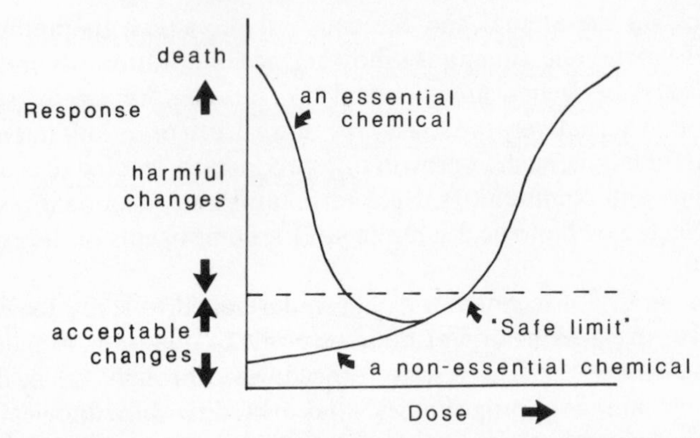

Figure 5.2 The response of a 'typical' population to changes in the dose of essential and non-essential chemicals. The position of the dotted line on the arbitrary vertical axis varies with the toxicity of the chemical, and for some poisons, e.g. carcinogens, the dotted line may not exist, there being a continuous gradient of response with concentration. The intercept of the dotted line with the response curve indicates the 'safe' dose limit. (After Pascoe, 1983.)

sub-lethal (or chronic) testing of pollutants as an alternative. Chronic testing may, for example, look at reproductive success or metabolic responses of animals subjected to low doses of a pollutant. The use of low-level, long-term testing is essential if we are to fully understand the problems caused by dilute concentrations of pollutants, however they are more difficult to administer, control and reproduce than acute tests. Thus, despite the drawbacks, acute testing remains the principal method for the laboratory assessment of pollutants, which lead to guidelines for the disposal of toxic materials.

All laboratory-based assessments of pollutants should be supported by comprehensive field-based studies of the actual effects of the pollutants on estuarine organisms. For a field assessment of the effects of a pollutant discharged into an estuary, the principal method is to collect samples of animals or plants from areas believed to be affected by a polluting discharge, and to compare them with samples from areas believed to be unaffected. Because they are stationary, and therefore more reliably sampled than mobile fauna, most sampling programmes collect benthic animals or attached algae. In a typical estuarine sampling programme, samples of the benthos from the intertidal or subtidal areas of an estuary would be collected, often from a spatial grid around a polluting source. The material would be returned to the laboratory, and all the animals and plants collected identified. A comparison would then be made between the estuarine community close to the source of pollution and those farther

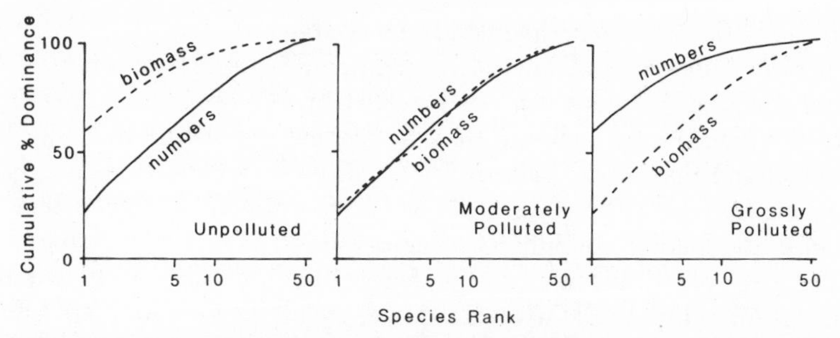

Figure 5.3 Hypothetical k-dominance curves for species biomass and number, showing unpolluted, moderately polluted and grossly polluted conditions. On the horizontal (x) axis, the species are ranked in order of importance, using a logarithmic scale, and on the vertical (y) axis, the percentage dominance is plotted, using a cumulative percentage scale. (After Warwick, 1986.)

away. Such comparisons can be made from the basis of the number of species collected, the abundance of individual animals, the biomass of the organisms, or by chemical analysis of the specimens.

A full description of field methods for sampling estuarine organisms is given in Baker and Wolff (1987), Morris (1983) and Holme and McIntyre (1986). Warwick (1986) has shown that the results of field studies may be used to assess the pollution status of a benthic community, by plotting the distribution of the numbers of individuals among species (Figure 5.3). In unpolluted conditions the benthic community will be dominated by one or a few large species, each represented by rather few individuals. In polluted situations, benthic communities become dominated numerically by one or a few small species, each represented by many individuals. Other methods of plotting such results will be discussed in section 5.4.

In any field assessment of the effects of pollutants within estuaries, due account should always be made for all environmental variables, natural as well as man-induced. For example, in a study of the effects of sewage sludge dumping in the Thames estuary, Talbot *et al.* (1982) found that the distribution of the benthic fauna was governed by the sediment type and degree of mobility of the sea bed. In one possibly polluted area the fauna was impoverished, and dominated by one polychaete, *Polydora ciliata*, and in another dumping area, the fauna was enriched. Overall it proved difficult to assess any effects attributable to dumped sewage in an area where there is great spatial variation both in sediment type and associated faunal groupings.

In order to assess the effects of disturbance from pollution, dredging or other activities, field experiments can be established in estuaries, whereby an area is deliberately disturbed or defaunated, and the recovery of the fauna observed. In such a study in Alewife Cove, Connecticut, USA, Zajac and Whitlach (1982a, b) found that the response to disturbance was very variable, and that no set pattern for estuarine re-colonisation succession could be detected. Thrush and Roper (1988) have also shown that re-colonisation studies are difficult to assess for pollution monitoring.

The assessment of stress and the degree of contamination of estuarine fauna at the sub-population level requires the use of common, representative indicator (or sentinel) organisms which can be used for bioassay or bioaccumulation studies. Suitable sentinel organisms need to be static or at least have only a limited home-range. Around the world the blue mussel, *Mytilus edulis*, has been widely used in 'mussel-watch' programmes, and amongst fish the flounder (*Platichthys flesus*) and the eel-pout (*Zoarces viviparus*) have been shown to be very suitable for estuarine studies (Elliott *et al.*, 1988).

All methods for detecting the effects of pollutants in estuaries are a compromise between the desirable, the practicable, the affordable and the reliable. It is generally true that almost any method will be able to detect severe sources of pollution, or 'hot-spots', but lesser sources of pollution require a wide variety of approaches, and the skill and ingenuity of the scientist may be severely tested. Such skill is especially needed for estuarine studies, since a wide range of natural environmental variables, such as salinity, temperature or sediments, occur alongside any perturbations induced by mankind's use and abuse of the estuary.

5.4 Organic enrichment

The sources of organic enrichment within estuaries are varied. Principal amongst these is domestic sewage, since many towns and cities situated on estuaries discharge their sewage untreated, or with only primary treatment (major solid material removed). Additionally, food processing industries, such as canneries, breweries or distilleries may discharge organic waste, and pulp and paper-producing industries can also be major dischargers of organic material, in the form of cellulose waste. The waste from agricultural industry ranges from ground water enriched with fertilisers, through to slurry produced from intensive animal rearing. Intensive mariculture for fin-fish or shellfish is a further source of organic enrichment.

The effects of all these various activities are similar, despite the differing origins of the organic matter. Organic matter is, of course, vital to the functioning of the estuarine ecosystem, as the basis for detritus food-chains. A small amount of organic matter, well dispersed, can be readily utilised within the estuarine ecosystem to enhance the levels of biological production. Problems arise, however, if a large volume of organic effluent is discharged into an estuary with a low flushing rate, or if too much effluent is discharged at a single point. When excess quantities of organic matter are present, then bacteria and other micro-organisms which utilise the organic matter will consume all the available oxygen in the water. Such problems may be particularly severe in summertime because of reduced river flow coupled with temperature increases which lead to enhanced bacterial activity, and thus accelerate the depletion of the oxygen content of the water.

The development of low oxygen, or even anoxic, conditions leads to the extinction of the normal macrofauna of the estuary, and its replacement by annelid worms, especially oligochaetes. In areas of greatest organic enrichment even such worms may be excluded. The microbial degradation of organic waste will, apart from consuming the available oxygen, also lead to increased sulphide production. The oxidation of sulphide leads to further oxygen depletion.

The effects of organic enrichment are most clearly seen by monitoring the fauna and measuring the redox potential (E_h) of the sediment at successive distances away from a known organic discharge. The redox potential may be determined with special platinum electrodes inserted into the sediment, with the resultant negative E_h values indicating reducing (anoxic) conditions, whilst oxygenated sediments have positive E_h values. Pearson and Stanley (1979) give full details of the measurement of E_h.

The results of the measurement of fauna and E_h at distances of up to 800 m from the discharge of organic effluent from an alginate (seaweed) factory are presented in Figure 5.4. Close to the effluent were relatively few species, with low numbers and biomass, coupled with negative E_h values. At 100 m the abundance of animals (mainly small polychaetes) increased dramatically, but there are still few species. The biomass and E_h rose more steadily. Between 100 and 400 m the total numbers (abundance) of animals fell, whilst the species diversity and E_h steadily increased. The biomass showed a dip at 200 m, but otherwise increased. Beyond 400 m the species diversity, abundance and biomass all fell, whilst the E_h continued to rise. Such patterns of change in the species diversity (S), total abundance of animals (A), and total biomass (B), have been recorded in the vicinity of many organic discharges, and the form of the curves have become known as

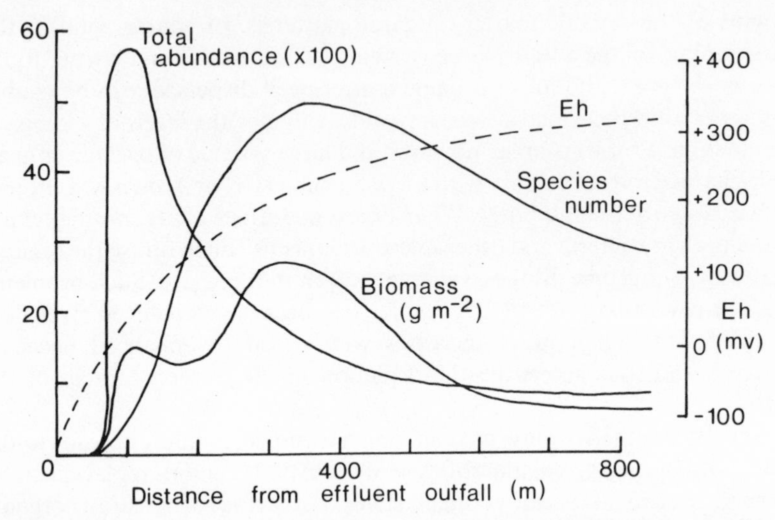

Figure 5.4 Species number, total abundance of animals, biomass and E_h along a gradient of decreasing sedimentary organic content in Loch Creran, Argyll, Scotland, expressed as distance from the effluent of a seaweed processing plant. (After Pearson and Stanley, 1979.)

'S–A–B curves' (Pearson and Rosenberg, 1978). Such S–A–B curves are now regarded as typical for the effects of organic enrichment, and can be interpreted as follows. Close to the effluent, conditions are anoxic, and no macrofauna can live. As one moves away from the effluent the sediment is colonised by 'opportunistic' animals, mostly small annelids, which occur in enormous numbers. This zone, the 'peak of the opportunists' has low species diversity and modest biomass. Beyond this is the 'ecotone', where the opportunists are replaced by the more normal fauna, and as shown at 200 m in Figure 5.4, a dip in the biomass may occur. Beyond the ecotone, is the 'transition zone' where the fauna shows enrichment due to the organic source, with maximal biomass and species diversity. Beyond this enriched zone the fauna declines to its background levels.

The change in species composition is shown pictorially in Figure 5.5. Water movement and renewal is an essential factor affecting the impact of organic material in a particular area. The importance of water renewal is shown in Figure 5.6. With regard to an estuary the importance of water renewal is such that a given organic discharge into the well-mixed mouth of an estuary may have little or no measurable effect, whereas the same discharge into the restricted confines of the head of the estuary may produce severe symptoms of organic enrichment.

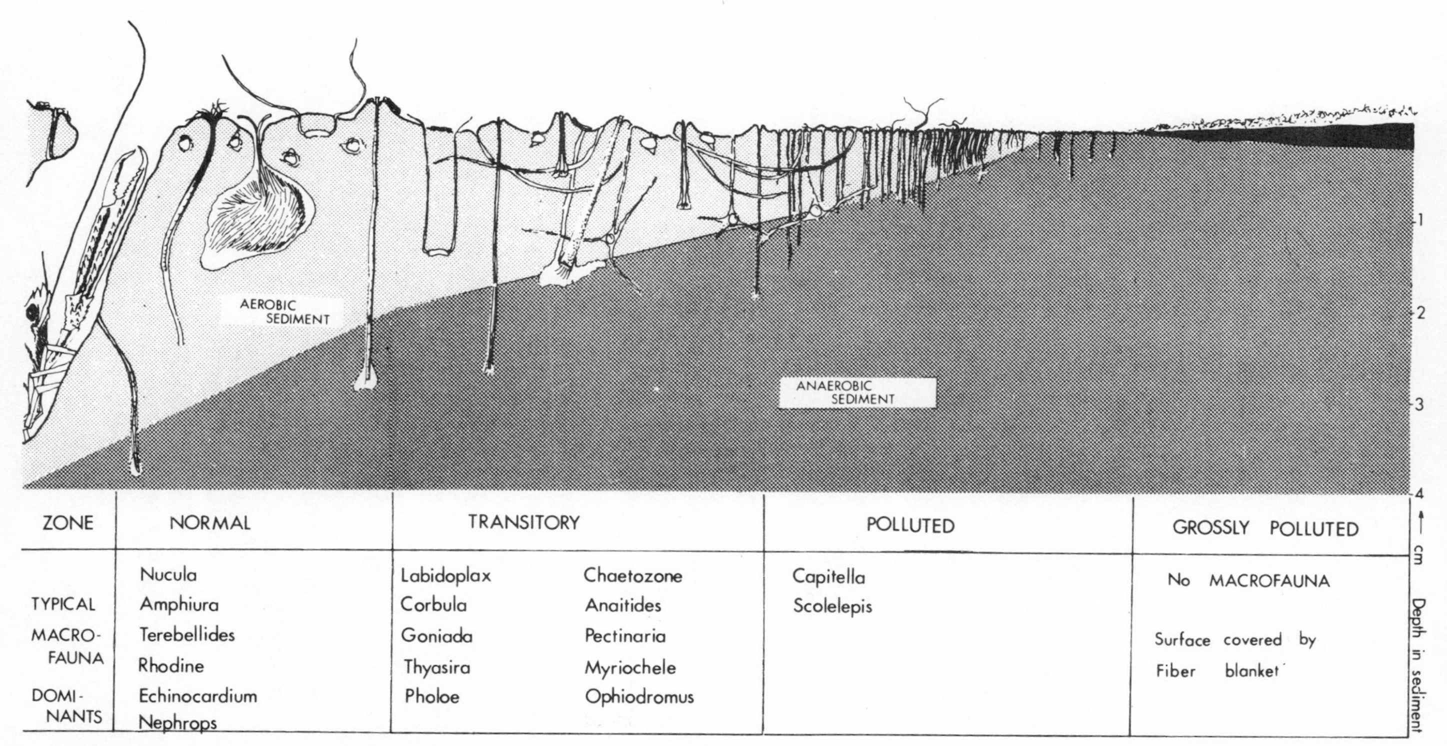

Figure 5.5 Diagram of the changes in fauna and sediment structure along a gradient of organic enrichment. The species listed are typical macrofaunal species for north European marine waters, and equivalent species will be found elsewhere. (From Pearson and Rosenberg, 1978.)

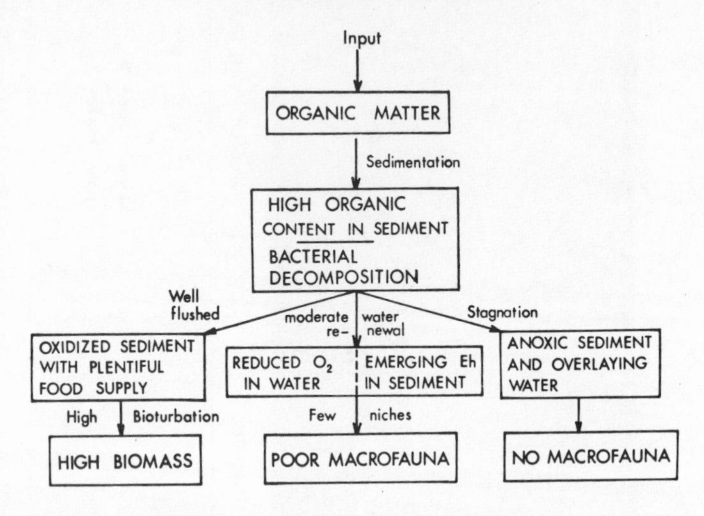

Figure 5.6 Diagram to show some pathways of organic input to the marine and estuarine environment, and its effects in relation to water renewal. (From Pearson and Rosenberg, 1978.)

In an alternative approach, Leppakoski (1975) classified the benthic fauna of brackish-water areas into five groups of species with reference to their response to organic pollution. These five groups are:

1. *Progressive species of the first order*
   Scarce in unpolluted areas, but occur in great numbers in polluted areas. Their numbers increase as the degree of pollution increases, until inhibitory effects are noticed in the most polluted areas.
   e.g. Oligochaeta—*Limnodrilus hoffmeisteri, Tubificoides benedeni*
   Polychaeta—*Polydora ciliata, Capitella capitata*
   The 'opportunists' of S–A–B curves.
2. *Progressive species of the second order*
   Tolerate only slight pollution, but increase in number as the degree of pollution increases.
   e.g. Polychaeta—*Nereis diversicolor, Manayunkia aestuarina*
   Mollusca—*Macoma balthica, Thyasira* spp.
   Includes many typical estuarine species.
3. *Regressive species of the first order*
   Tolerate only slight pollution, but decrease in number as the degree of pollution increases.

e.g. Crustacea—*Corophium volutator, Gammarus* spp.
Mollusca—*Hydrobia* spp., *Cerastoderma* spp.
The majority of benthic species are in this group.

4. *Regressive species of the second order*
Clean water species, totally absent from polluted areas.
e.g. Crustacea—*Pontoporeia affinis*, Ostracoda
Few estuarine species in this group.

5. *Indifferent species*
Distribution unaffected by pollution.
e.g. Annelida—*Tubifex costatus, Glycera alba.*

The particular species listed here are based on studies in Sweden and Finland, and need not occur elsewhere. However, the division into 'progressive' and 'regressive' species with regard to pollution can be made in any estuary in the world.

Figure 5.7 Map of the Wadden Sea. Stretching from Texel in The Netherlands to Esbjerg in Denmark this is the largest estuarine area in Europe. (After Essink, 1978.)

The Wadden Sea is the largest estuarine area in Europe, occupying about 10 000 km^2 and is the principal haunt of millions of shore birds as well as being a key nursery area for the fishes of the North Sea (Figure 5.7). Organic pollution of the Wadden Sea reflects the distribution of populations and industry, with untreated or minimally treated effluent from dairies and slaughterhouses in the Danish sector, effluent from the estuaries of the Elbe and Weser in the German sector, and from major Dutch cities. Viewing the Wadden Sea as a whole the organic effluents cause severe local problems of oxygen deficiency, but on a gross scale the area remains as probably the most important and productive estuarine ecosystem in Europe.

One estuary which has recovered dramatically from decades of organic pollution is the Thames estuary which flows through London, UK. Due mainly to massive amounts of domestic sewage the state of the Thames declined steadily during the first half of the twentieth century, until by the 1950s it was completely anoxic over the middle reaches, inhabited only by oligochaete worms, and mallard and swans which fed on grain spillages. Conditions only began to improve in 1964 with the completion of a new sewage works at Crossness, and improved further in 1974 due to another sewage works at Beckton, coupled with the closure of several old discharges. As a result of these improvements the condition of the estuary water has steadily improved, and by 1975 the lowest oxygen content was 25% of saturation. Since then the water quality has steadily improved, and as it has done so, fish and birds have returned to the estuary. Figure 5.8 records the number of fish species caught during the recovery period. In the period 1920–1960 no fish were caught. The eel was the first to return, and by 1975 36 species were found, including most dramatically the return of salmon, *Salmo salar* (Andrews and Rickard, 1980). A wide diversity of invertebrate fauna now occurs, although there are fewer oligochaetes. As the oligochaetes have reduced so have the numbers of tufted duck and pochard which fed upon them. The numbers of wading birds and other ducks have increased dramatically in reponse to the return of a wide diversity of prey organisms. Adjacent to the Thames, the Medway estuary has also shown an increase in fish diversity and abundance as organic pollution has abated (Wharfe *et al.*, 1984).

The fish populations in the Clyde estuary, which flows through Glasgow, Scotland, UK were eliminated by 1845–1850 due to industrialisation and urbanisation and the consequent organic waste. Considerable effort has been made to reduce this organic pollution, which has resulted in 18 fish species being recorded in 1978, 25 in 1980, 32 in 1982 and 34 in 1984.

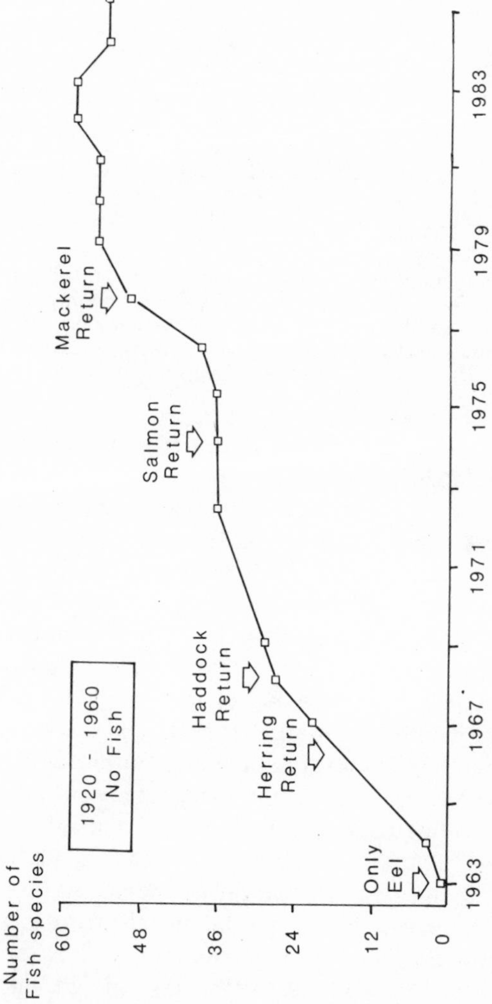

Figure 5.8 Number of fish species recorded in the Thames estuary, England. Drawn from data supplied by Thames Water.

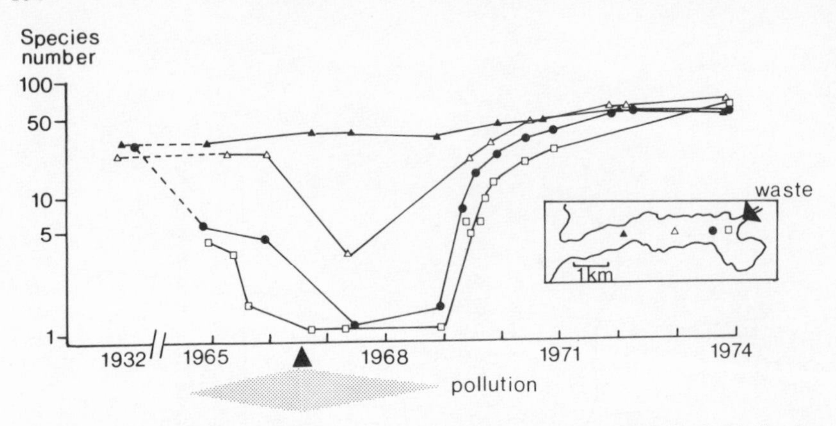

Figure 5.9 The number of benthic species present at four stations in Saltkallerfjord, Sweden between 1932 and 1974. The positions of the four stations are shown on the insert map. Waste material from a pulp mill was discharged to the fjord in increasing amounts until July 1966. From that date the amount discharged has declined considerably with a consequent increase in species diversity. (After Rosenberg, 1976.)

Biological recovery is not complete in this system, but it can be seen to be progressing well (Henderson and Hamilton, 1986).

The recovery of the benthos in an organically enriched estuary, the Saltkallerfjord, has been traced by Rosenberg (1976). From his data (Figure 5.9) it can be seen that in 1966 there were few animal species near to the waste discharge from a sulphite paper mill, but over 30 species at 4 km distance. Following closure of the mill in that year the number of species at all stations has increased, although a lag of over 3 years was observed before recovery became clear. By 1974 there were approximately 50 species at all stations, irrespective of distance from the old discharge.

The 1980s have seen the rapid growth of mariculture in many countries. In Scotland, Norway and Canada the main expansion has been in the rearing of salmon in floating cages (Figure 5.10). The salmon are usually fed on pelleted food, composed mainly of fish meal, and up to 20% of this food may not be intercepted and will fall to the bottom, along with the faeces produced by the fish (Figure 5.11). Since the fish cages are usually placed in shallow waters, the waste material rapidly falls to the bottom and accumulates there. As Brown *et al.* (1987) have shown, underneath the floating salmon cages the sediment is highly reducing (negative E_h) and azoic. A highly enriched zone, dominated by *Capitella capitata* and *Scololepis fuliginosa*, occurred from the edge of the cages out to approximately 8 m. A slightly enriched zone occurred at up to 25 m, beyond which

Figure 5.10 A floating sea-cage fish-farm for salmonid fish (salmon or trout). Such farms have become common sights on the fjordic estuaries of Norway, Ireland, Canada, New Zealand, Scotland and elsewhere. (See also Figure 5.11.)

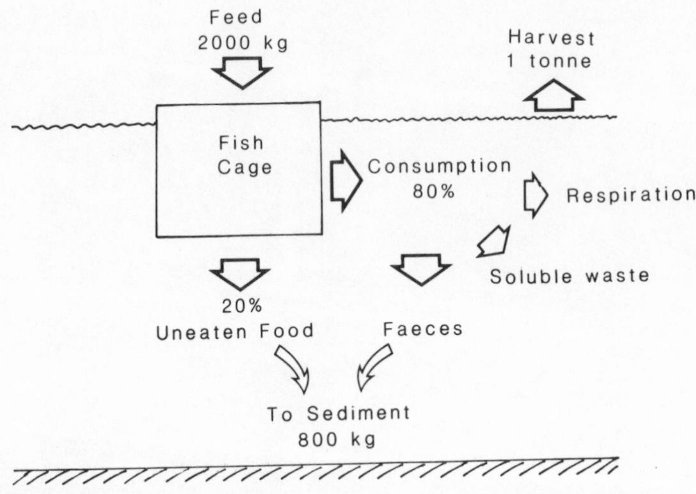

Figure 5.11 The environmental impact of the floating-cage culture of salmonid fish. Values are expressed in terms relating to the production of 1 tonne of fish. (After Gowen *et al.*, 1988.)

the fauna was unaffected by the cages. Thus the ecological effects of this form of organic enrichment are severe but limited.

In Spain, New Zealand and Ireland, farming of mussels has expanded. The mussels are grown on suspended ropes, feeding on natural planktonic material. In studies of mussel farming in the estuarine Ria de Arosa of north-west Spain, Tenore *et al.* (1982, 1985) have shown that mussels are a 'key species' in determining ecosystem structure and dynamics of the whole area. By intensive mariculture, man has replaced the zooplankton with mussels as the dominant herbivores in the area. The major changes are:

1. The surface area of, and detritus from, the mussels support a dense epifaunal community, which utilises 90% of the mussel faeces, and supplies food to demersal fish and crabs
2. Epifaunal larvae, rather than copepods, dominate the zooplankton community
3. Nutrient cycling by mussels dampens phytoplankton oscillations and contributes to high seaweed production on ropes
4. Heavy sedimentation of mussel deposits changes the sediment regime and lowers infaunal production
5. Transport of particulate organics derived from mussel deposits from the farming areas enhances benthic biomass outside this area, and might support near-shore fisheries, especially for hake.

Tenore concluded that the raft culture of mussels affected food-chain patterns and production in generally positive ways, although admitting that the infaunal benthos near the farms was typical of polluted conditions. Considering this study along with several others, we can summarise in Figure 5.12 the possible effects of mussel farming on the estuarine ecosystem. While mussel rafts or ropes reroute the flow of energy or materials, they do not add any extra nutrients to the ecosystem, unlike caged fin-fish fed on prepared food, and any changes involve processes different from eutrophication caused by an enhanced nutrient supply.

To reduce the effects of organic waste discharges into estuaries, sewage works are constructed. Whether providing primary treatment (the removal of solids only) or secondary treatment (full biological treatment of liquid waste as well), considerable volumes of solid waste, sewage sludge are produced. The most economical method of disposing of this sludge is to transport the waste by ship from the sewage-works to a dumping site at the mouth of the estuary, or further offshore. Two types of dumping site can be distinguished, either accumulation sites where the material accumulates on the bottom, or dispersal sites where currents disperse the material over a wide area. A major sludge dumping site is in New York Bight, where large quantities of waste from New York City have been dumped near Ambrose Light. The bottom water oxygen concentrations are depressed at and around the dump site, and in spring and summer 1976 an area of 12 000 km^2 experienced oxygen concentrations less than 2 mg l^{-1}, with a smaller area becoming completely anoxic, resulting in the death of at least 143 000 tonnes of the clam *Spisula*. Reporting the situation in 1980–82, Steimle (1985) has shown that there is not an azoic 'biological desert' in the dumping area, but the benthic fauna shows enhanced production of the stress-tolerant species, leading to an enhanced benthic biomass (127–344 g m^{-2} wet weight) with productive demersal fisheries.

The dump site for the sewage of the city of Glasgow is at Garroch Head, where 1.5 million tonnes of sewage sludge are dumped annually. This is a typical accumulating site, and the biomass and abundance of the fauna is very high at the centre of the dump site (being 500 g m^{-2}, and about 500 000 individuals m^{-2}), although the species diversity is low with the principal species being nematodes (*Pontonema* spp.), the oligochaete *Tubificoides benedeni* and the polychaetes *Capitella capitata* and *Scololepis fuliginosa* (Pearson *et al.*, 1986). As Figure 5.13 shows, the species diversity increases away from the epicentre, whilst the biomass and abundance decline. These changes in species diversity, abundance and biomass conform well to the general pattern described above.

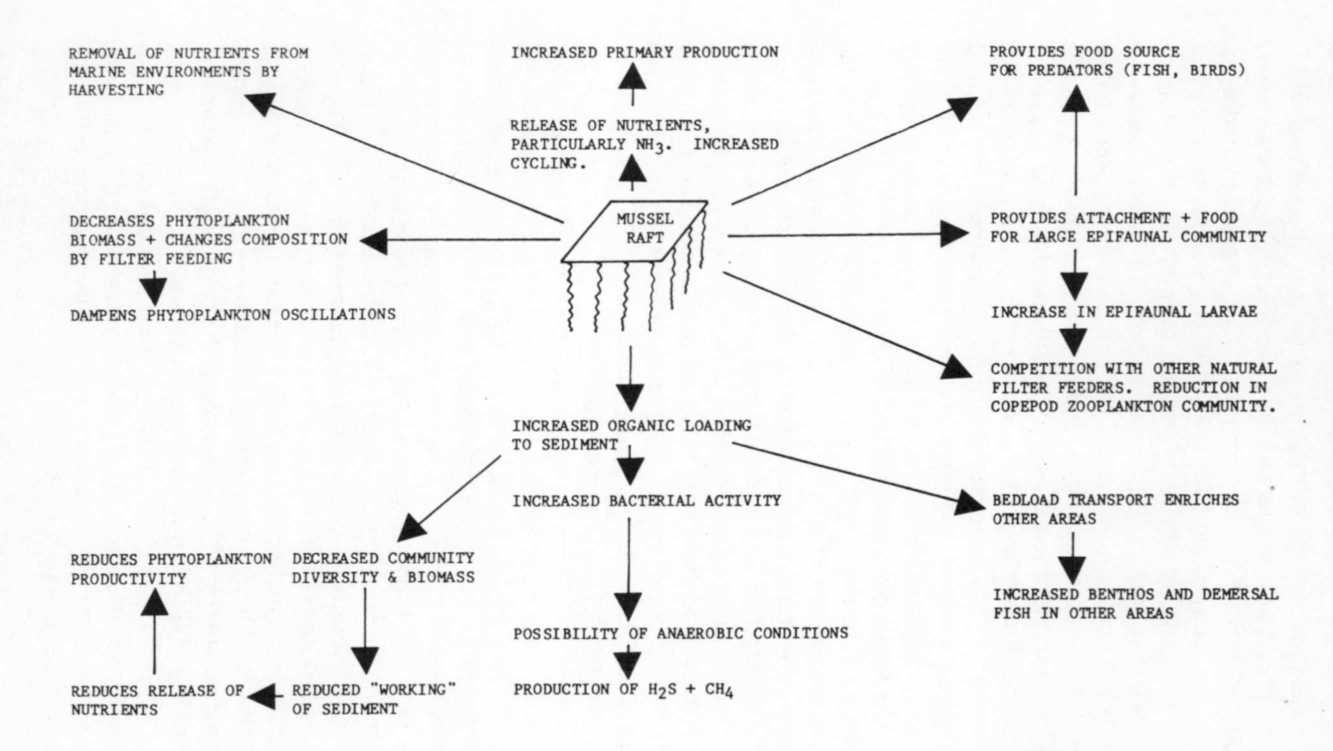

Figure 5.12 Summary of the possible effects of mussel farming. Note that some of the effects are contradictory, and not all effects will be seen at one site. (From Gowen *et al.*, 1988.)

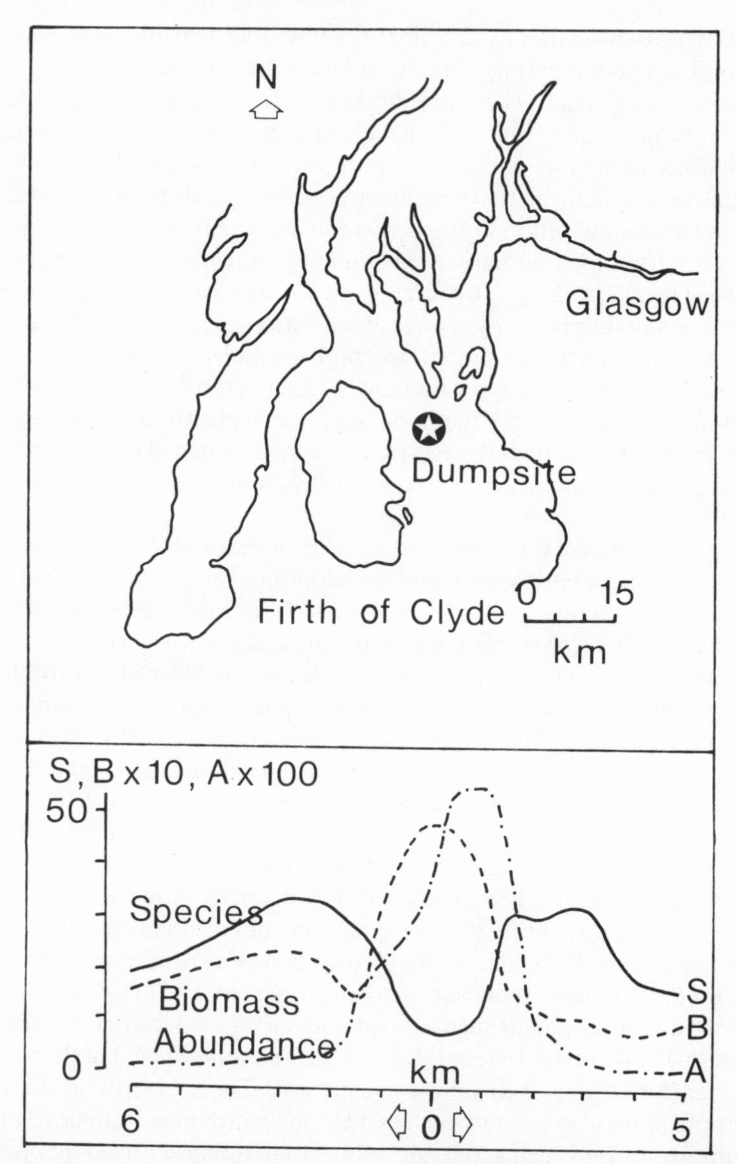

Figure 5.13 Location of the Garroch Head sewage sludge disposal site in the outer Clyde estuary, Scotland, in upper diagram. The lower diagram shows the biomass (B, g wet weight m^{-2}), total abundance (A, m^{-2}) and species number (S) of the benthic infauna along a 11 km east/west transect across the disposal site. (After Pearson, Ansell and Robb, 1986.)

The sewage from the city of Edinburgh (up to 500 000 tonnes annually) is dumped at two alternating sites, in summer near Bell Rock and in winter near St. Abb's Head. These are dispersing sites and Moore and Davies (1987), in monitoring the sites, have found no evidence of any organic carbon accumulation in the sediments, and only slight changes in the benthic fauna, showing that the sludge is sufficiently dispersed to preclude any significant alteration to the benthic environment. Five million tonnes of sewage from the London area are dumped annually on a dispersing site in the Thames estuary. However, concern has been expressed at the desirability of dumping any sewage at sea, and even if it is fully dispersed a large amount of organic matter has still been added to the ecosystem. In defending the marine disposal option, Mackay (1986) claims that due to problems with alternative methods, such as incineration or disposal on land, the marine option is the best practical environmental option, and that an accumulating site which can be accurately monitored is preferable to a dispersing site.

The addition of extra organic matter to estuaries and the sea is generally known as organic enrichment, and the addition of extra nutrients from the same source is known as *hypernutrification*. Hypernutrification can stimulate the growth of phytoplankton in the process known as *eutrophication*. In fresh waters it is the supply of phosphates which may cause eutrophication, but in estuaries and the sea it is usually the supply of nitrogen which is the critical factor. The effects of eutrophication are usually observed as increased densities of green phytoplankton and benthic micro- and macroalgae, decreased transparency in surface waters and/or oxygen deficiency at the bottom, sometimes associated with mortality of the bottom fauna and bottom-dwelling fish (Rosenberg, 1984, 1985).

Eutrophication has been known to be occurring in the Baltic Sea for some time. During the 1980s, the symptoms of eutrophication have also been found in the Kattegat, the shallow sea between Denmark and Sweden, with marked changes in the bottom fauna observed, and mass mortality of bottom-dwelling animals such as *Nephrops norvegicus* (scampi, or Norway lobster). The demersal fisheries (for cod, etc.) have declined, but the pelagic fisheries (for herring, etc.) show some increase. During eutrophication the phytoplankton may produce an 'algal bloom' which may be toxic to other organisms. In 1987 such a 'red tide' bloom of toxic algae forced the closure of 170 miles (250 km) of clam and oyster beds in North Carolina. Eutrophication has now been recorded in many estuaries of the world, as an example, the steady increase of phytoplankton in the Ems–Dollard estuary in response to eutrophication, measured here as total phosphates, is

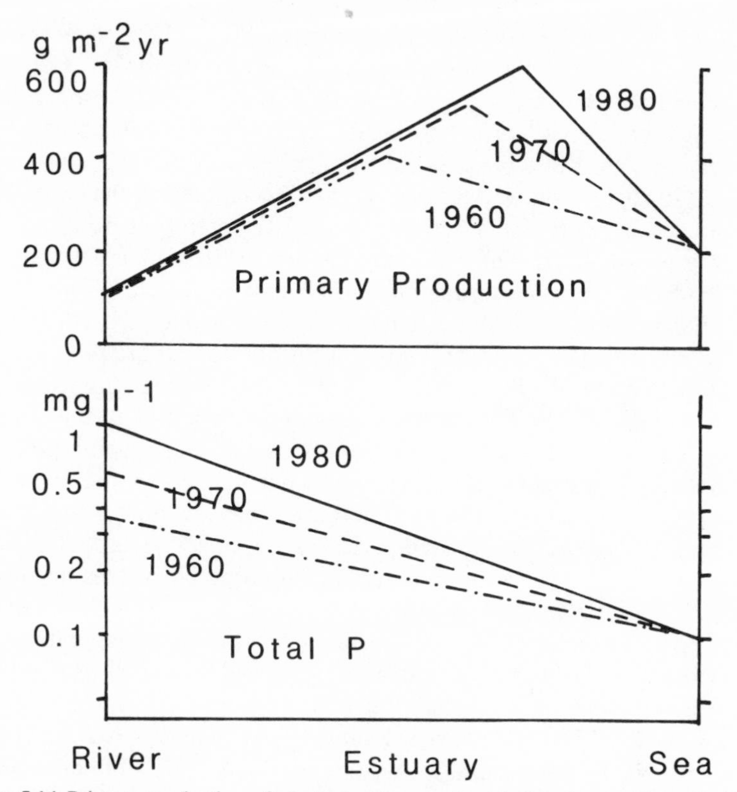

Figure 5.14 Primary production of phytoplankton, and total phosphorus (P) in the Ems-Dollard estuary in 1960, 1970 and 1980. Note the eutrophication related to increased supply of P from river and inland sources. (After Postma, 1985.)

shown in Figure 5.14. On the intertidal mudflats of many estuaries, the growth of extensive 'algal mats' is a clear symptom of eutrophication. Algal mats of *Enteromorpha* or *Cladophora* will smother the primary consumers in the benthos, killing animals such as the bivalves *Cerastoderma* or *Mya*, but increasing grazers such as *Hydrobia*.

Although sources such as sewage, mariculture waste, or sludge dumping do contribute some extra nutrients to estuaries, it has been widely shown that the major source of extra nitrogen is from rivers, due to drainage from the fertilisers used in agriculture, and from inland sewage works. It has been generally estimated that less than half of the fertiliser applied to agricultural crops becomes bound into the harvested crop, the remainder is lost through drainage into rivers and thus enters estuaries. From fields adjacent to estuaries it will enter directly. Atmospheric deposition of nitrogen is the second most important source, after fertilisers. The supply of agricultural nutrients to estuaries does stimulate plant production in estuaries, but not all of the nutrients are or can be utilised within the estuary, and the nutrients pass out from the estuary into the adjacent sea. Within Europe, it is now clear that the whole ecosystem of the North Sea is threatened by such excess nutrients which are entering it from the many estuaries bordering it. These nutrients are derived from inland sources, and carried from the rivers, through the estuaries into the sea. Without doubt the biggest source of such nutrients is the river Rhine.

The consequences of organic enrichment of estuaries are thus shown to be variable, and all the skill of the professional biologist is required to interpret their effects. The understanding of the operation of the estuarine ecosystem lies at the heart of our comprehension of organic pollution. Organic enrichment in small amounts may produce 'acceptable' changes in estuarine systems, whereas large amounts may produce 'catastrophic' changes resulting in mass mortality. As Van Impe (1985) has clearly shown, a long-term increase in the numbers of estuarine birds does not always appear to be an indication of an improvement of the environmental quality of an estuary, it may in fact indicate an increase in estuarine pollution. In the Scheldt estuary an incontestable ecological deterioration, due to several sources of mainly organic pollution, has led to a greater food supply for intertidal birds. The birds have responded by visiting the area in increased numbers, and by staying longer. In contrast to other forms of pollution, the effects of organic pollution are usually reversible, so the estuarine ecosystem with its in-built flexibility can accommodate a moderate amount of organic pollution, as well as recover fairly rapidly from severe organic pollution.

5.5 Industrial contamination

In this section we shall consider the waste from industrial sources which enters estuaries. Much of it enters estuaries from industries located on the banks of the estuary, but it should also be remembered that waste from inland sites may be discharged into rivers or public sewers which subsequently arrive in estuaries. The waste from industrial sources may be treated before discharge, so that the effluent arriving in the estuary has a considerably reduced concentration of waste materials, or the waste may be discharged in an untreated form.

Because of the attraction of flat land for development, coupled with the proximity of suitable harbours and other transport systems for the import and export of raw materials and finished products, estuaries throughout the world have become sites for many major industries, especially the oil and chemical industries, the electricity generation industry, the shipbuilding industry, and metal-producing industries such as aluminium smelting. All these industries produce some amount of waste, ranging from large volumes of water which have been used for cooling purposes, through to chemical waste products which may be extremely toxic even in small quantities.

Crude oil is probably the most complicated natural mixture on earth, although dominated by hydrocarbons, it also contains non-hydrocarbons and metals. Carlberg (1980) has estimated that over 2 million tonnes per annum is discharged into estuaries, being approximately one-third of all the oil discharged into the world's oceans. Thus estuaries do indeed receive a disproportionately large burden. Most of this oil enters estuaries from river run-off, marine transportation, or domestic or industrial waste.

The effects of the oil industry on estuaries may be divided into, (1) the impact of gross spillages, due to shipping accidents or human error at a loading terminal, and (2) the impact of effluents produced by refinery and petrochemical industries. In the case of oil spillages from harbours or collisions the effects may be varied, since oil contains many different fractions, some of which are toxic whilst others are relatively inert. When oil is spilt from a harbour or a collision it reaches water as a separate phase, which is largely immiscible with the water, and generally forms a surface slick. As the slick spreads it will undergo evaporation, and many of the lighter and most toxic elements can be lost by weathering, but unfortunately, in the confined space of an estuary there may not be time for this to occur before the spilt oil is deposited on the shore. On the shore the oil acts mechanically, smothering animals in burrows or on rocks, and excluding the light from plants. On a salt marsh the soiled parts of the plant die, but if it can survive this, the plant can regrow. The water-soluble or more volatile compounds are most toxic, and spillages in American estuaries of fuel-oil containing 45% low aromatic oils have devastated commercial shell fisheries. The heavier compounds within the oil, which remain after evaporation, will tend to sink, where they undergo microbial decomposition by many micro-organisms which are capable of degrading petroleum hydrocarbons as well as naturally occurring biogenous hydrocarbons.

The commonest advice on how to deal with oil spills in estuaries is to contain the oil by a boom if possible, and then to use mechanical removal with pumps for the trapped oil. If it comes ashore, mechanical removal with shovels or excavators is the preferred mode of treatment, rather than emulsifiers if at all possible. The earlier emulsifiers, or detergents, used in clean-up campaigns to disperse oil were often more toxic to estuarine life than the original oil. Newer emulsifiers are less toxic. Figure 5.15 shows the decision processes involved in dealing with an oil spill within an estuary, as recommended by CONCAWE.

Less dramatic, but potentially more dangerous to estuaries than oil-spills, are the insidious effects of the continual discharge of industrial effluents from petrochemical complexes. The effluent from such industries is mostly hot fresh water, which may contain certain amounts of chemicals, including hydrocarbons (oil) from refineries, or chemical waste products such as phenols or ammonia from chemical works. A refinery effluent may contain only 10 to 20 parts per million of oil which is difficult both to see and to remove, but in a flow of 455 000 l minute^{-1} this represents 6825 l day^{-1} of oil (Nelson-Smith, 1972). The impact of effluent from petrochemical complexes in the Medway and Forth estuaries has been examined by Wharfe (1977) and McLusky (1981). In the vicinity of the outfall there is generally an abiotic zone, with no life at all. Beyond this is a 'grossly polluted zone' in which only oligochaetes are found, in small numbers. Then follows a 'polluted zone' with abundant oligochaetes and *Manayunkia* (a small polychaete), and occasional specimens of other species such as *Hydrobia*. Finally comes the 'largely unpolluted zone' at over 1 km from the effluent, with *Macoma, Cerastoderma, Nereis, Nephthys* and *Hydrobia* often abundant, and fewer oligochaetes. It may be seen that the zonation of species and their abundance is remarkably similar to that noted above (section 5.4) for the effects of organic enrichment. Since oil is a biological product derived geochemically from organic material, it is perhaps not so surprising that the effects of oil on estuarine ecosystems are rather similar to the effects of other excess organic materials. Throughout the world oil refineries have made strenuous efforts to reduce their waste effluent, and modern oil refineries have considerably less impact on the estuarine ecosystem than their older counterparts. When an oil refinery discharge is reduced, or terminated, then the resilient estuarine fauna can generally recolonise the area rapidly.

The effects of other chemical waste products, such as toxic chemicals or heavy metals, which may be included in petrochemical waste, may be quite different to the effects of hydrocarbons. Particular attention has been paid to discharges of heavy metals such as mercury, cadmium, copper, iron,

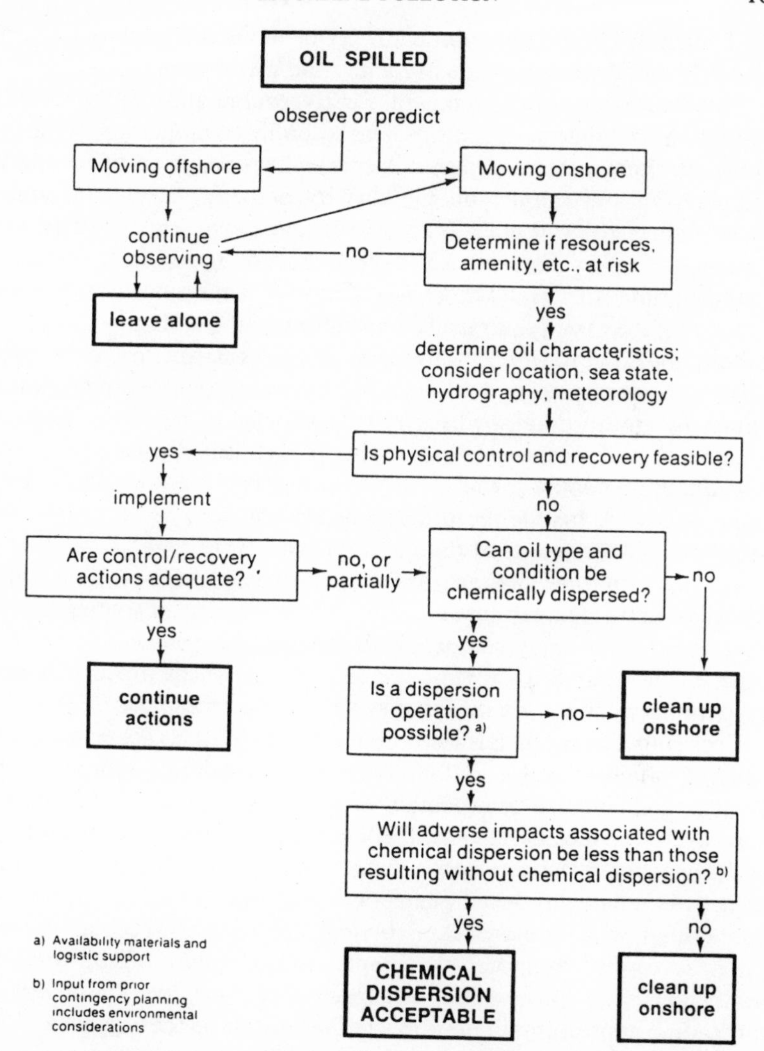

Figure 5.15 Decision 'tree' for the response to an oil spill in an estuary, considering in particular whether a chemical dispersant is acceptable. (From Concawe, 1981.)

lead, zinc, or chromium, from a variety of industries including chemical plants, metal-processing industry and mine waste water.

Two approaches have been used. Firstly, examination of the toxicity of the metals to estuarine organisms, and secondly, examination of the metal concentrations within organisms. A comprehensive review of the effects of temperature and salinity on the toxicity of heavy metals to estuarine invertebrates by McLusky *et al.* (1986) has shown that toxicity values determined under fixed (or single) temperature and salinity regimes are inappropriate for evaluating the effects of environmental factors in modifying the toxicity of metals to estuarine animals (Figures 5.16, 5.17). Indeed, the range of toxicity values for a single estuarine species exposed to different temperatures and salinities may exceed the entire range of toxicity values previously published for a wide variety of marine or freshwater animals. A rank-ordering of the toxicity of metals is: mercury (most toxic) > cadmium > copper > zinc > chromium > nickel > lead and arsenic (least toxic). A taxonomic order may also be seen of: Annelida (most sensitive) > Crustacea > Mollusca (least sensitive).

For all estuarine animals, heavy metal toxicity increases as salinity decreases and as temperature increases. From a variety of studies it appears that heavy metals may compete with calcium and magnesium at cellular uptake sites and thus disrupt the key physiological process of osmoregulation which is so vital for survival in estuaries.

Concentrations of metals below those needed to cause acute toxicity may cause a variety of sub-lethal effects, such as inhibiting reproduction or growth, and will also be accumulated within the tissues of the organisms exposed to the metals. The metal content will generally increase with size and age as the metals are accumulated, and unable to be excreted. Bryan *et al.* (1985) have produced guidelines for the assessment of heavy metal contamination in estuaries using biological indicators, such as seaweeds, molluscs or fish as monitors. Simple measurements of the total concentration of metals within estuarine organisms may however, give misleading impressions of the impact of the metals on the organism. Within estuarine molluscs, Coombs and George (1978) have shown the formation of vesicles within cells which enclose the metal within a membrane. These vesicles prevent contact of excess metal with vital constituents and effectively detoxify the metal. Estuarine annelids, such as *Nereis* have been shown to be able to detoxify metals by accumulating them within their jaws. Although detoxification may prevent a mollusc or an annelid from being adversely affected by excess metals, a predator eating it may not possess the same ability and may be affected by the total concentration of

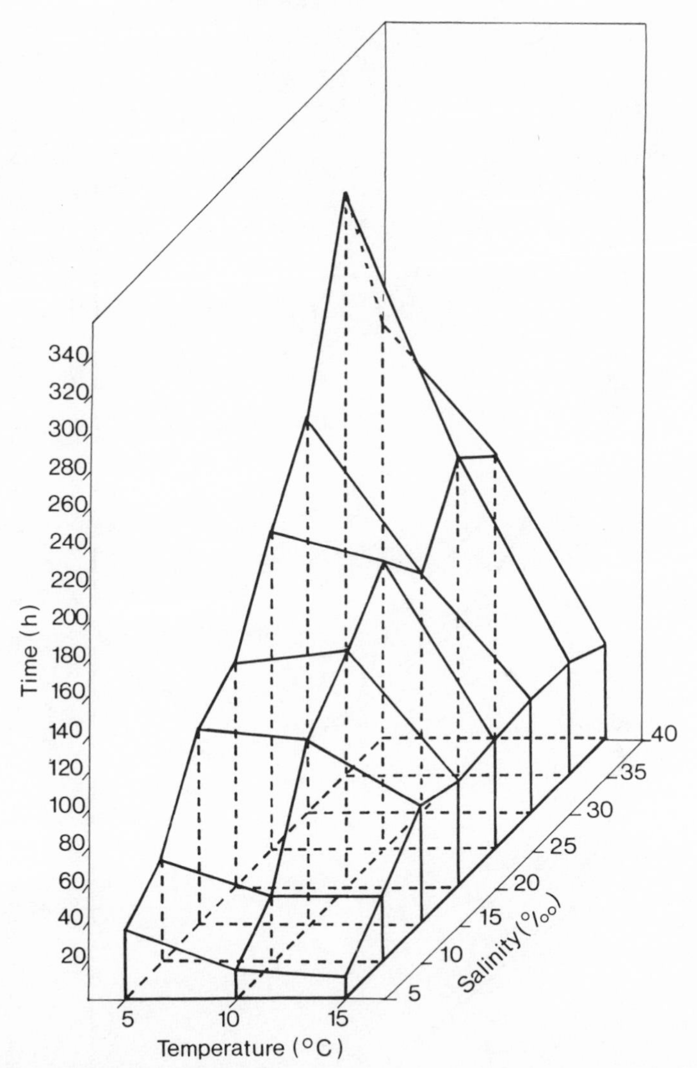

Figure 5.16 The effect of temperature and salinity on median survival time (h) of *Corophium volutator* at a chromium concentration of $16\,mg\,l^{-1}$. (After Bryant *et al.*, 1984.)

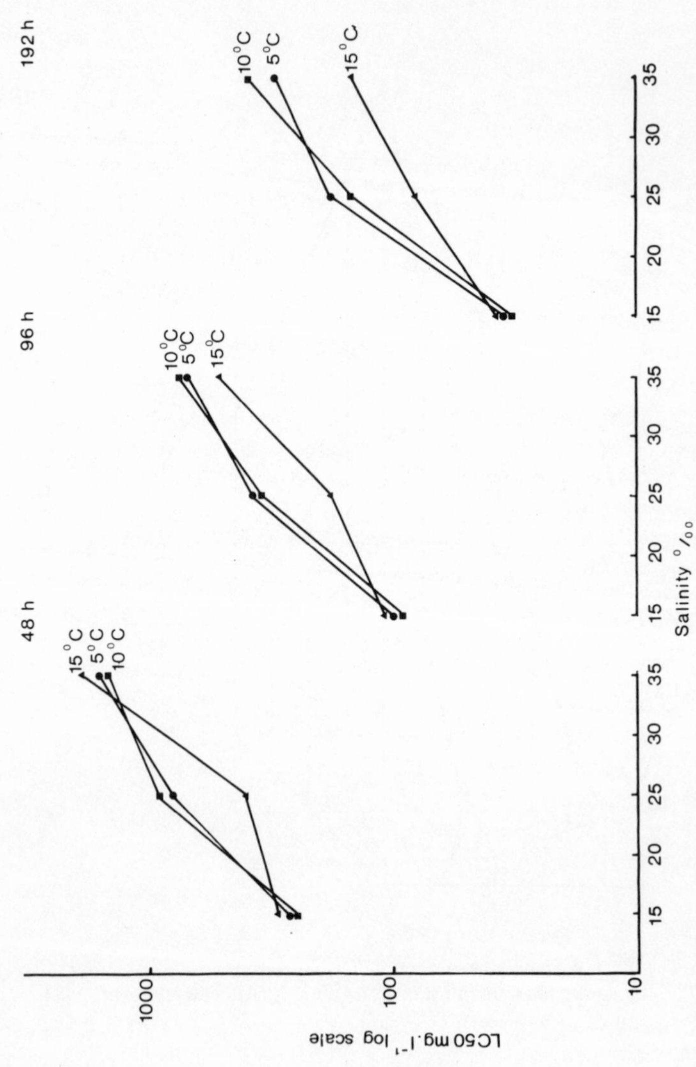

Figure 5.17 The effect of temperature and salinity on the median lethal concentration (LC_{50}) of nickel to *Macoma balthica*. (From McLusky *et al.*, 1986.)

the metal. Unlike some other pollutants, heavy metals may show *biomagnification*. That is, they show an increase in concentration as one proceeds up the trophic levels of the estuarine ecosystem. For that reason, humans should always avoid eating any shellfish exposed to heavy metals.

The processes of sedimentation within estuaries, discussed in the first chapter, may serve to relieve the estuarine and marine ecosystem of some of the worst effects of heavy metal pollution. As metals are discharged into the turbid waters of estuaries they may become rapidly bound onto the surface of the fine sedimentary particles. As the sedimentary particles settle on to the intertidal mudflats, the metals are gradually buried. In many of the estuaries bordering the North Sea, it has been estimated that about half of the metals entering the estuary, become trapped within the estuarine sediments, and only a lesser amount is eventually discharged to the sea. However if estuaries are to serve as such vital traps for pollutants, it is essential that the natural intertidal mudflat areas remain intact and undisturbed. In several cases it has been shown that man's activities may disturb such metal burial, for example dredging will often liberate heavy metals from sediments, and even bait-digging for worms has been shown to lead to an increase in the total concentration and bioavailability of metals (Howell, 1985).

The addition of pesticides and organohalogen compounds to estuaries poses considerable threats to the estuarine organisms. Unlike petroleum hydrocarbons these materials, the halogenated hydrocarbons (hydrocarbons containing chlorine, bromine, iodine or fluorine), are not degraded by chemical oxidation or micro-organisms. Thus, like metals, these substances are permanent additions to the estuarine ecosystem, but, compared to metals, even minute quantities of these man-made substances can pose a hazard to estuarine life. The organohalogen compounds are the PCBs (polychlorinated biphenyls), including HCB (hexacholorobenzene), HCH (lindane, hexachlorocyclohexane), dieldrin, and DDT (dichloro-diphenyl-trichloroethane) used both as pesticides, and for industrial purposes (especially in electrical equipment). These substances cannot be excreted by any animals which ingest them, and they thus accumulate within their bodies, principally in fatty tissue. This bioaccumulation leads up through the food-chain, by biomagnification, so that top predators have the highest concentrations, and may suffer most. Until the effect of these substances was realised, and their production curtailed, about 1 million tonnes of PCBs were produced around the world. Much of this material has ended up in estuarine and marine organisms, and will continue to circulate through ecosystems.

Pesticide or industrial PCBs in very low concentrations e.g. 1 $\mu g\ l^{-1}$ (= 1 part per billion) have been shown to be lethal to estuarine life both in laboratory tests, and in field mortalities around industrial and argicultural discharges containing pesticides. Due to the biomagnification effects of PCBs, top predators such as birds and seals are particularly at risk. Within the Baltic Sea, for example, high levels of PCBs within seals have been held responsible for the failure of reproduction in many female seals.

The subject of the addition of radioactive substances to estuarine environments arouses considerable public concern. Man has, during the latter part of the twentieth century, added radioactive material to sea waters from the fallout of atmospheric bomb testing, from waste discharged from nuclear power stations and reprocessing plants, and from the use of radioactive material in submarines. The unique detectability of radioactive isotopes enables the anthropogenic additions to the environment to be clearly identified and quantified. However it is often forgotten that the man-made additions may be very small in comparison to the amount of radioactivity which naturally occurs in sea water. Over 90% of the natural radioactivity of sea water arises from potassium-40, almost 1% from rubidium-87, and the remainder is virtually all from the isotopes of uranium. The concentrations of these elements vary proportionately with salinity, so the natural radioactivity of estuarine water is governed by salinity, with greater concentrations in saline waters and lesser concentrations in brackish waters. At a typical salinity of 33‰, the natural radioactivity has been measured as 12 Bq/l (One Bq (Bequerel) is one disintegration per second). In the Forth estuary and the Firth, the volume of water is 20 km^3 or 2×10^{13} l, so Leatherland (1987) has calculated the total natural radioactivity of these waters as 240 000 GBq, indicating a total of 60 t of dissolved uranium in these waters. In this area a naval dockyard adds 7 GBq per annum, and low levels of isotopes will be added in sewage systems derived from hospital or laboratory sources. A nuclear power station planned for the area will add a further 200 GBq per annum, but it can be seen that these sources are minute compared to the natural levels of radioactivity present.

In Britain and France the major anthropogenic source of radioactivity is from nuclear reprocessing plants at Sellafield and La Hague. These plants recover uranium for re-use from the fuel rods of nuclear power stations, and discharge large quantities of waste water with a low radioactive content, but because of the volumes involved they generate a considerably greater input of radioactive material than nuclear power stations or other manmade sources. Some of the waste products decay rapidly or are adsorbed onto fine sediments. The fine sediments can be carried into local estuaries, and in the case of Sellafield these need to be regularly monitored. The isotope caesium-137 remains in solution, and since it does not occur naturally it serves as an effective 'marker' for the Sellafield waste. Regular annual surveys, such as Hunt (1988), of British coastal waters, show that waste from Sellafield is predominantly carried north and west away from the discharge, and after passing around the north of Scotland enters the North Sea and its estuaries. Close to the discharge point the concentration of caesium-137 is 0.5 Bq kg^{-1}, and declines to 0.05 Bq kg^{-1} in the North Sea (Figure 5.18). These concentrations are considerably less than the natural radioactivity levels of sea water, but are greater than any other anthropogenic sources of radioactivity. In recent years the quantities of waste discharged from reprocessing plants has diminished considerably, but they remain the main source of manmade radioactivity in estuaries. Laboratory studies have shown acute lethal effects of radiation doses on many organisms (Clark, 1986), but it must be said that existing levels of radiation in estuaries and the sea have so far produced no measurable environmental impact on these ecosystems.

Of the various pollutants discussed, three groups of substances pose particular threats to the estuarine ecosystem. These are heavy metals, PCBs and radioactivity. All these substances are persistent, and cannot be degraded, although they show some slow reversibility due to burial in sediments. All show bioaccumulation, and in the case of PCBs also show biomagnification. All show effects on the fauna at low concentrations, often at concentrations well below those shown to be toxic in standard 96-hour toxicity tests. The threats posed by metals and PCBs can be demonstrated in present estuarine ecosystems, but radioactivity effects have not been shown to disrupt ecosystems so far. By contrast, organic materials and oil may show severe effects in the vicinity of discharges, but these materials can be biologically degraded, and their effects can be completely reversible at the estuarine ecosystem level. So, when considering pollutants in estuaries one must consider not only the quantity and quality of the waste, but also whether it can be degraded and whether its effects are reversible.

5.6 Reclamation and engineering works

Estuaries are areas of natural deposition of sediments, and as salt marshes develop they slowly create new areas of dry land. These new areas of dry

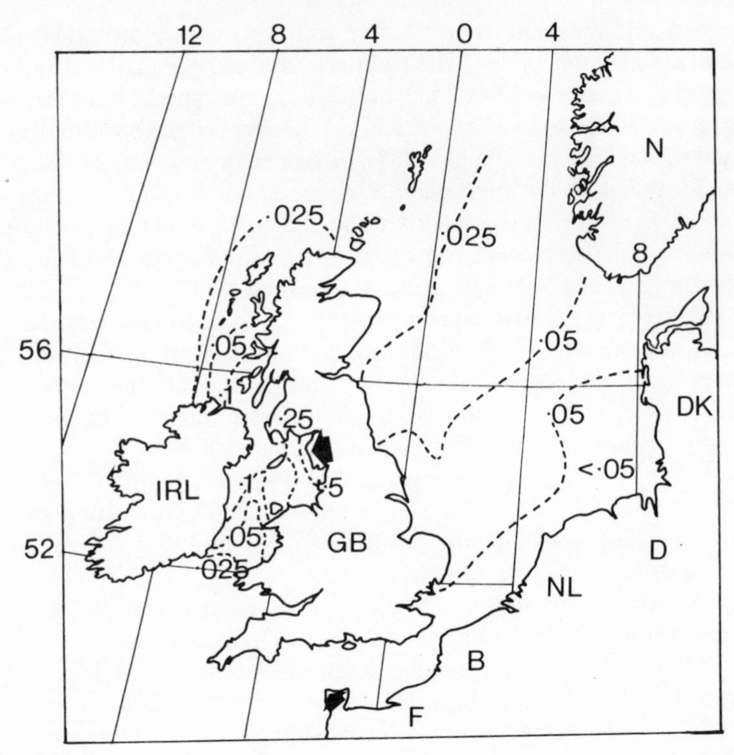

Figure 5.18 Concentration (Bq kg^{-1}) of caesium-137 in filtered water from north-east Atlantic waters, 1987. The arrows show the discharge points of material from nuclear reprocessing plants. (Based on data in Hunt, 1988.)

land are often acquired by neighbouring farmers seeking to extend their acreage. Throughout the world, for many centuries, man has drained salt marshes and with the protection of sea walls converted them into productive agricultural land. In Britain such reclamation has been predominantly on the east coast, and in Denmark such reclamation has occurred in West Jutland. In India and Bangladesh such reclaimed land has proved to be particularly suitable for rice cultivation. Such processes of agricultural reclamation are, however, gradual in comparison to the large-scale reclamation schemes undertaken to meet the needs of industry. Modern industrial developments seek areas of flat land close to available transportation, and as other land may not be readily available, large-scale estuarine reclamation is often suggested. For such a land development scheme, the first stage is usually the construction of a sea-wall or bund across the intertidal mudflats, followed by the infilling of the bounded area with dredged estuarine mud, or hard-fill material derived from quarries, mine waste, ash waste from coal-fired power stations, or domestic refuse. In other schemes, the impounded area may be filled with fresh water to form reservoirs, and especially in the Netherlands the impounded area may be drained and converted into agricultural land or polders.

Whatever the use of the reclaimed land, it represents a total loss to the estuarine ecosystem. In the estuaries of eastern Britain, for example, industrial reclamation has obliterated over 90% of the Tees estuary, leaving less than 10% of the original intertidal area (Figures 5.19, 5.20). On

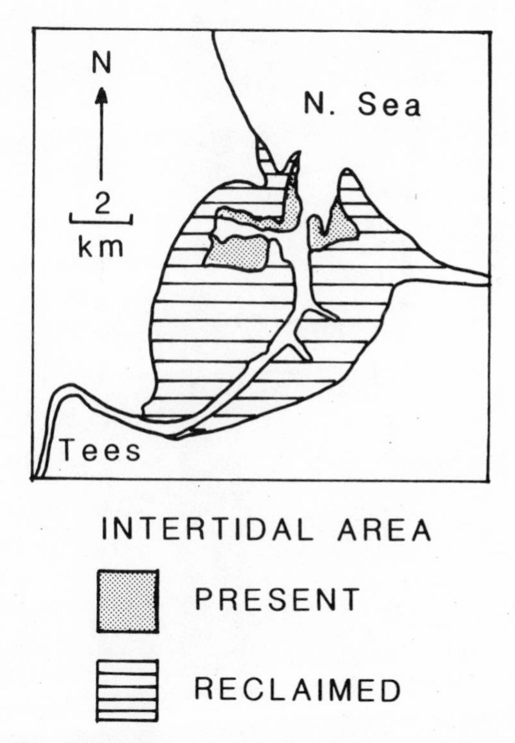

Figure 5.19 Reclamation of intertidal areas of the Tees estuary in north-east England for industrial purposes over the last 60 years. (After Carter, 1988.)

Figure 5.20 The Tees estuary, UK, where, like many other industrialised estuaries, much of the intertidal habitat has been reclaimed to provide sites for industry and associated docks. This view shows the transporter bridge across the docks, in an area where the estuary has been almost completely canalised. (See also Figure 5.19.)

the Forth estuary approximately 50% of the intertidal area has been reclaimed in the past 200 years (Figure 5.21). At first such 'reclamations' were for agricultural purposes, followed by harbour works, and more recently for industrial purposes, including sites for power stations (Figure 2.9), refuse disposal and fly-ash dumping. Within Southampton Water reclamation has taken place over many years to meet the needs of new docks, oil refineries, power stations and other industries (Figure 5.22). Whilst the salinity regime has been hardly affected by the reclamation schemes, the removal of extensive salt marshes has reduced substantially the input of organic detritus to the ecosystem, and made the whole ecosystem more dependent on phytoplankton production. As intertidal areas have been obliterated, the commercial clam harvest has fallen from 100 t to 60 t per annum, and the bird population feeding intertidally has declined substantially.

San Francisco Bay is the largest estuary on the Pacific coast of the USA, with a water surface area of 1240 km^2, and probably the most modified estuary in America. Large areas of wetland, or salt marsh, have been destroyed by reclamation, so that only 6% of the original 2200 km^2 of wetland remain. Due to inland water engineering schemes less than 40% of the original river flow now reaches the estuary, a factor which has considerably reduced the flushing of pollutants from the Bay area. On the east coast of the USA more than 25% of the estuarine wetlands have already been lost through reclamation. Wherever, and for whatever purposes, reclamation occurs, the birds and fish which feed on estuarine mudflats simply find that a large proportion of their food supply has disappeared. American legislation now requires that new reclamation schemes must compensate for the loss of habitat by the provision of alternative localities; however, it is very doubtful if such compensation can ever fully replace a habitat which has taken centuries to develop.

The use of estuaries for transportation requires harbour works which can range in size from a single pier to a vast container or bulk-cargo terminal. A single pier may cause localised change in current pattern and sedimentation, whereas a large terminal will have a similar impact to other major reclamation schemes. In order to maintain access for shipping, many estuaries are dredged. At the dredge site, the substrate is removed along with any fauna it contains, and recovery of the infaunal benthos may take up to 10 years. The disturbance of dredging will also increase the turbidity of the water, which may affect areas outside the dredge site. The dredge spoil is transported to a spoil ground, usually at the mouth of the estuary, or outside it. The fauna of the spoil ground is obviously affected by being

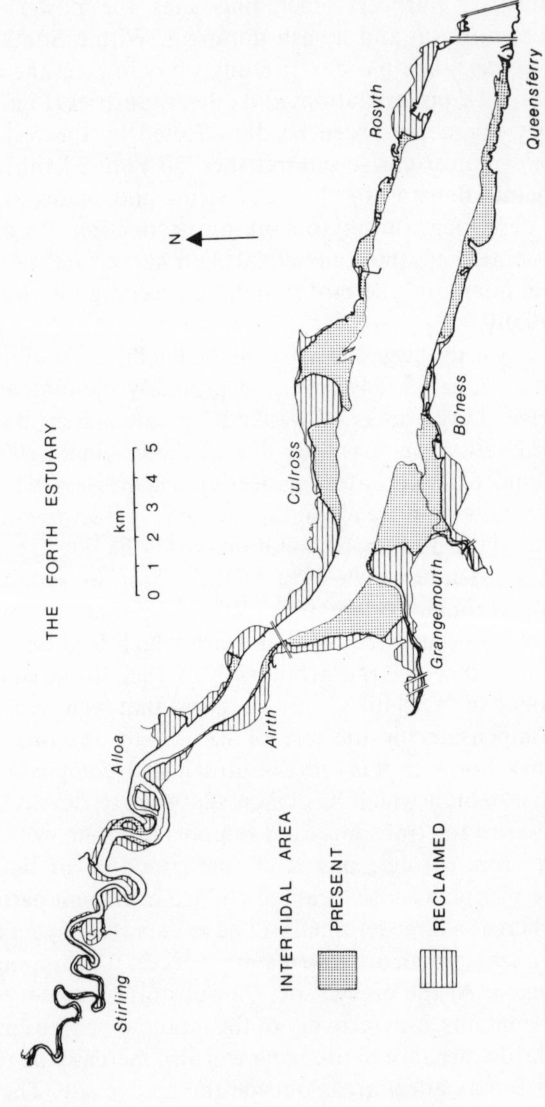

Figure 5.21 Reclamation of intertidal areas of the Forth estuary in Scotland for agricultural and industrial purposes over the last 200 years. (From McLusky, 1987.)

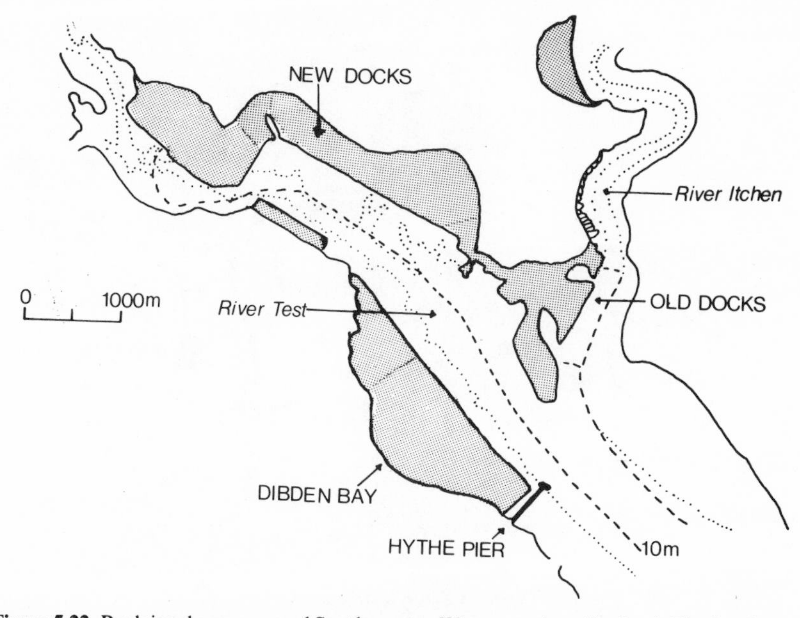

Figure 5.22 Reclaimed areas around Southampton Water, southern England. The fine dotted line represents the low water mark of 1830, and the firm dotted line the current dredged channel. (After Coughlan, 1979.)

smothered, and increasingly it has become realised that pollutants such as heavy metals which adhere to sediment particles are, instead of being deposited and buried within estuaries, liberated into the water by dredging activity and transportation to the spoil ground. Harbour sites may also be a cause of pollution, especially from oil, through accidents as ships are loaded and unloaded, or due to the risk of collision, or from leaks from ships' engines or bilges. Ship repair and demolition sites also present potential hazards for considerable further 'accidental' pollution.

Faced with the problem of the threat of large-scale flooding from the North Sea, the Netherlands have reclaimed extensive estuarine areas, such as the Zuider Zee, which now forms the freshwater Ijselmeer as well as agricultural polders. In the past two decades major changes have occurred in the estuaries of the southern Netherlands, in the 'Delta Plan', whereby the estuaries of Rhine, Meuse and Scheldt have been dammed up and converted from tidal estuaries into freshwater lakes, brackish lakes, or sea areas with restricted tides (Figure 5.23). The southernmost estuary, the Westerscheldt, remains open, but the Oosterscheldt estuary has been cut off

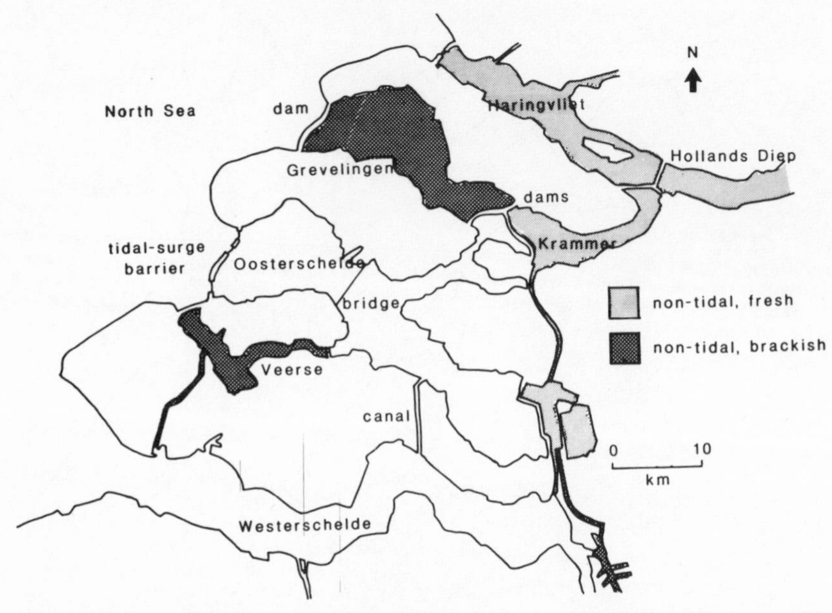

Figure 5.23 The Delta scheme area of south-west Netherlands. The map shows that four previously open estuaries have been partially or completely closed off from the sea. In addition inland compartment dams have created freshwater lakes from previous estuaries. Only the Westerschelde remains as an estuary. (After Rijkswaterstaat, 1986.)

from its freshwater supply (the Rhine/Meuse), large areas of shallow wetland have been reclaimed, and an 8 km wide tidal storm surge barrier restricts the inflow of sea water. As a consequence it is now a sheltered marine bay, not an estuary. The Grevelingen estuary has been closed at both its river and seaward ends to form a semi-stagnant non-tidal brackish-water lake, with a stable salinity of about 18‰. The small Veere estuary has also become a brackish-water lake with a salinity of 7‰ in the winter, rising to 12‰ in the summer. The Haringvliet estuary has been closed at its mouth, preventing the entry of salt water, but still permitting the exit of freshwater. Consequently it has become a freshwater lake, fed by the Rhine. The necessity for the Delta Plan to the safety of the Netherlands is clear, but the consequence has often been the obliteration of mussel beds and the elimination of feeding sites for birds and fish.

All down the east coast of America, the level of the sea is rising and the intertidal areas are in retreat. Sea level has risen by a global average of about 12 cm over the past 100 years, due to global heating which has caused glaciers and ice caps to melt and thus increased the total volume of the world's oceans. Increased global warming over the next 25 years is expected to raise the sea level by another 10 cm. Estimates by the US National Research Council suggest rises in sea level of 0.5–1.5 m by the year 2100 (Hecht, 1988). In addition to rises in sea level, the level of the land may change. In recently glaciated areas such as Alaska, Scotland or Finland, the land is rising, with rates of rise of 14–19 mm per year recorded in Alaska, and 3 mm per year in Scotland. On the eastern seaboard of USA, the land is sinking at 1.5 mm per year in New York, and up to 12 mm per year in Texas and Louisiana. In London, and southern England, land is sinking at about 2 mm per year. The subsidence of the land, coupled with the rise in sea level, means that the sea is eroding shorelines. Particularly vulnerable are the almost flat intertidal areas and wetlands of estuaries, and it has been estimated that Louisiana is losing about 40 hectares of wetlands every day, or about 15 500 hectares per year. In London, England, a tidal barrier has had to be built to protect it against high spring tides which, with the subsidence in land level and rise in sea levels, threatened to flood this major city. Elsewhere new sea walls will have to be built increasingly higher to protect coastal communities. For the estuarine ecosystem, these changes are yet another environmental factor which will change many of the estuaries and wetlands that we see around us today.

In order to provide electricity in the next century, as conventional sources such as oil, gas, coal or nuclear power become scarcer or less environmentally acceptable, various schemes have proposed the development of tidal power barrages across estuaries. In France, a tidal power station on the Rance estuary has operated successfully for 20 years. In Canada the tides on the Bay of Fundy have an amplitude of 12 m, and a small tidal power station is now operating at Annapolis Royal, Nova Scotia, with plans being prepared for more major stations in the future. China and the USSR also have small stations in operation, whilst they prepare for larger schemes. In Britain plans are well advanced for tidal power stations on the Mersey and Severn estuaries, with the latter able to supply 20% of Britain's electricity requirements. In all cases, the tidal amplitude of the estuary above the power station barrage is reduced, leading to loss of intertidal habitat as the distance between high and low water is reduced. Such an environmental change would reduce the feeding areas available to birds which feed intertidally. Other likely changes are to sedimentation or water quality, as currents are reduced and 'flushing' diminished.

Against the ecological losses of estuarine habitat caused by reclamation may be set the occasional gains of lagoons or other wetlands, but more typically reclamation represents only gains to agricultural or industrial land users, and losses to the wildlife and fisheries of estuaries. For many estuaries the destruction of habitat due to various reclamation schemes has had a far greater impact than any of the effects of any polluting discharges entering the estuary. For the sea as a whole the increase in nutrients, due to excess agricultural fertilisers, entering from estuaries, which received them from rivers, poses a major long-term threat.

In this chapter we have seen the many ways in which mankind uses and abuses the estuaries of the world. In the next chapter we will consider the management practices which may lead to controlling the effluents discharged to estuaries, or limiting the areas which experience reclamation, and also the information which is required to make such decisions, and can ultimately lead to national and international laws to protect the estuarine ecosystem.

CHAPTER SIX
THE MANAGEMENT OF ESTUARIES

6.1 Introduction

The main objective of estuarine management is to devise a framework within which man may live harmoniously with nature. Estuarine management can be divided into three broad areas; policy, planning and practice (Carter, 1988). Policy relates to the political and administrative framework through which estuarine management is regulated, be it by legislation or education. Planning is the process of resource allocation, be it by ecological or economic yardsticks. Practice covers the techniques needed for the implementation of planning decisions, such as building new sewage-works, or treating industrial effluent.

The management of estuaries has to cope with three fundamental paradoxes. Firstly, the majority of the world's major cities are located alongside estuaries, yet, for most of the inhabitants of those cities, estuaries are the most natural wildlife habitat that they encounter. Secondly, most of the major estuaries of the world are to some degree polluted, yet, in many countries there are more estuarine nature reserves (or sites of special scientific interest in the UK) than for any other habitat. Thirdly, many estuaries are organically enriched, yet all estuaries are amongst the most productive natural ecosystems known.

6.2 Policies

In adopting a policy for estuarine management different approaches can be adopted. Either, an Environmental Quality Objective (EQO) policy, usually linked to an Environmental Quality Standard (EQS) policy, or a Uniform Emission Standard (UES) policy. The essence of the EQO approach is that an estuary (or any water body) can disperse, degrade and assimilate pollutants which enter it. It is based on the concept of 'assimilation capacity', which is the capacity of a body of water to cope with the effluents discharged into it. Apart from any chemicals which bioaccumulate, it assumes that 'the solution to pollution is dilution!'. The EQO approach is concerned to define the overall conditions of the habitat and then to select for each estuary a set of permissible operations, or levels of pollutants, discharged in order to achieve that condition. The Environmental Quality Standards (EQS) are the levels of permitted discharges which are selected for a particular estuary. Thus the EQOs are the stated aims for an estuary, whereas the EQSs are the conditions used to bring about those EQOs. The alternative UES approach is to establish a national or international standard for the discharge of a pollutant, and then to apply those standards to every discharge, irrespective of the location of the discharge. The emphasis in a UES policy is clearly on the uniformity of discharge from different sources or locations.

US state and federal policy has generally followed the EQO/EQS approach. The UK government has also hitherto followed an EQO/EQS approach, but with the adoption of the North Sea declaration following a conference in 1987, it has moved to a UES policy for certain substances. These substances, on a so-called 'red list', whose control is a priority, will be controlled by cutbacks in all emissions irrespective of the body of water to which they are discharged.

A typical set of Environmental Quality Objectives would be that the quality of the estuary should allow:

1. The protection of all existing defined uses of the estuary system.
2. The ability to support on the bottom the biota necessary for sustaining sea fisheries.
3. The ability to allow the passage of migratory fish at all stages of the tide.
4. Low chemical and microbial contamination of the biota, which should not affect its consumption by man or other oganisms.

The above list has been chosen mainly on biological grounds, but there could be an EQO for each defined usage of the estuary. For example, water contact sports might require a different list (Sayers, 1986).

The uses which are recognised in the first EQO are: disposal of effluent, commercial fishing and angling, nature conservation, bathing, boating and water skiing, tourism and amenity, navigation, and water abstraction for industrial or agricultural purposes. All of these uses, except the disposal of effluent, require the estuary to be as free of pollution as possible, so it can be deduced that in order to meet that objective the disposal of effluent must be controlled, either by the reduction in volume, or by an improvement in quality of the waste so that it does not impede the other uses. At the mouth of the estuary, where there is plenty of water, this objective might be attained by better dispersion of existing waste through the construction of a long sea outfall waste pipe. But in the inner estuary, or in the rivers flowing into the estuary where there is less dilution water available, the attainment of the same objective might involve the construction of new sewage-works or treatment plants. Thus the standard of effluent permitted may be different in different parts of the estuary. In order to achieve the defined Environmental Quality Objectives, a higher Environmental Quality Standard would generally have to be applied to a sewage or industrial discharge at the head of an estuary, than to one at its mouth.

The second objective requires that the estuary be of sufficient quality to sustain the diverse components of the estuarine ecosystem that we have described in earlier chapters. In particular, no discharge to the estuary should obliterate the fauna in the vicinity of the discharge, and reclamation and other engineering schemes which destroy the benthic habitat should be avoided.

The third objective, to allow the passage of migratory fish such as salmon or eels at all states of the tide, is designed to ensure that the oxygen concentration, as a measure of water quality, is greater than 50% saturation at both high and low tide, and in both summer and winter. In many estuaries, this objective is readily attainable in winter at high tides, when there is a large volume of water, but more difficult to attain in summer at low tide, when there is a smaller volume of water and there will be less oxygen present due to the water being warmer or more saline (see for example Figure 6.3). This objective must therefore be defined in terms of the most problematic season or condition, not the most lenient.

The fourth objective is designed to eliminate contamination of, for example, shellfish by microbes, such as faecal coliform bacteria, which may be injurious to any person eating the shellfish. It also aims to eliminate contamination of fauna by any persistent accumulative pollutant, such as heavy metals (especially mercury or cadmium) or organohalogens which could taint them, affect predation, or harm anyone eating them.

The adoption of EQOs has been critised by several agencies and governments as being unjust. The opponents of EQOs say that they unfairly penalise a town or industry situated inland or near the head of an estuary, where there is little scope for dilution of an effluent, against a town or industry situated on a sea coast, or near the mouth of an estuary, where abundant scope for dilution is available. Thus more lenient EQSs might be applied to the discharge from an industry at the estuary mouth. To overcome this criticism, a set of numerical standards can be adopted, to which all industries or towns must conform. These standards, known as Uniform Emission Standards, or UESs, are applied directly to all effluents irrespective of location, in which case they are clearly different from EQOs and EQSs which relate to the quality of the receiving waters. UESs are particularly suitable for situations on a continent where a river may pass through several countries or states, such as the Rhine or the Danube, so that an industry in one country is not unfairly treated compared to similar industries in other countries.

Within Europe, the Commission of the European Community has been particularly active in issuing directives for the control of pollution, based on the UES approach, and has in recent years issued a series of standards through their directives. The directives are community law, but each needs a nation's own law to enable them to be adopted, and the nation has the choice of using the UES or EQS approach.

Some EQSs for estuaries and seas in Europe are listed in Table 6.1. Such lists are ultimately based on the evidence of the toxicity of the different substances, as discussed in Chapter 5, with mercury clearly regarded as the most toxic heavy metal. Such standards for EQS of waterborne contaminants may not necessarily reflect the potential harm to the fauna in the sediments. For example, mercury goes out of solution very quickly as it is adsorbed onto suspended particles, which then accumulate in sediments. There is therefore a move towards future EQSs being based on sediment

Table 6.1 Environmental Quality Standards (EQS) for Europe. All standards expressed as $\mu g\,l^{-1}$ (= parts per billion), in receiving waters (dissolved), except where indicated. After FRPB (pers. comm.) and Sayers (1986).

Parameter	*Estuary*	*Coastal sea*
Dissolved oxygen	>55%	
pH	6.0–8.5	
Ammonia	$<0.025\,mg\,l^{-1}$	
Cadmium	<5.0	<2.5
Copper	<5.0	<5.0
Mercury	<0.5	<0.3
Arsenic	<25	<25
Nickel	<30	<30
Lead	<25	<25
Zinc	<40	<40
Chromium	<15	<15
Iron	$<1.0\,mg\,l^{-1}$	$<1.0\,mg\,l^{-1}$

Table 6.2 The black list. Substances which should be banned or substantially eliminated from discharges. The list is 'List I' of the 'Paris' Convention, and European Dangerous Substances Directive.

List I contains certain individual substances which belong to the following families and groups of substances, selected mainly on the basis of their toxicity, persistence and bioaccumulation, with the exception of those which are biologically harmless or which are rapidly converted into substances which are biologically harmless.

1. Organohalogen compounds and substances which may form such compounds in the aquatic environment
2. Organophosphorus compounds
3. Organotin compounds
4. Substances in respect of which it has been proved that they possess carcinogenic properties in or via the aquatic environment (*)
5. Mercury and its compounds
6. Cadmium and its compounds
7. Persistent mineral oils and hydrocarbons of petroleum origin

and (for the purposes of implementing Articles 2, 8, 9 and 14 of this Directive):

8. Persistent synthetic substances which may float, remain in suspension or sink and which may interfere with any use of the waters

(*) Where certain substances in List II are carcinogens, they are included in category 4 of this list.

concentrations (Elliott, pers. comm.). It should also be remembered, as discussed in Chapter 5, that the toxicity of any single compound may vary greatly according to environmental factors within the estuary such as temperature and salinity, or may vary when the single compound is part of a cocktail of many chemicals.

In all international policies to regulate the discharge of toxic substances to marine and estuarine waters, two levels or scales of contaminants are recognised. Firstly a 'black' list of substances which should be banned or substantially eliminated from discharges, and secondly a 'grey' list of substances the discharge of which may be permitted in carefully controlled quantities. The composition of the lists shown in Tables 6.2 and 6.3 is based on Annexes I and II of the 'Paris' Convention which has been adopted by many countries, to control discharges into the seas from land-based sources. It matches closely the 'London' and 'Oslo' Conventions which are intended to prevent marine pollution by the dumping of waste and other materials at sea. The 'London' Convention covers the whole globe, and the 'Oslo' Convention covers the North Sea and north-east Atlantic. As a signatory to the 'London' Convention, the USA has adopted the substance of these lists. The black list is also List I of the European Dangerous

Table 6.3 The grey list. Substances whose discharge may be permitted in carefully controlled quantities. The list is 'List II' of the 'Paris' Convention, and the European Dangerous Substances Directive.

List II contains: substances belonging to the families and groups of substances in List I for which the limit values (referred to in Article 6 of the Directive) have not been determined; certain individual substances and categories of substances belonging to the families and groups of substances listed below; and those which have a deleterious effect on the aquatic environment, which can, however, be confined to a given area and which depend on the characteristics and location of the water into which they are discharged.

Families and groups of substances referred to in the second indent

1. The following metalloids and metals and their compounds:

1. zinc	6. selenium	11. tin	16. vanadium
2. copper	7. arsenic	12. barium	17. cobalt
3. nickel	8. antimony	13. beryllium	18. thalium
4. chromium	9. molybdenum	14. boron	19. tellurium
5. lead	10. titanium	15. uranium	20. silver

2. Biocides and their derivatives not appearing in List I
3. Substances which have a deleterious effect on the taste and/or smell of the products for human consumption derived from the aquatic environment, and compounds liable to give rise to such substances in water
4. Toxic or persistent organic compounds of silicon, and substances which may give rise to such compounds in water, excluding those which are biologically harmless or are rapidly converted in water into harmless substances
5. Inorganic compounds of phosphorus and elemental phosphorus
6. Non-persistent mineral oils and hydrocarbons of petroleum origin
7. Cyanides, fluorides
8. Substances which have an adverse effect on the oxygen balance, particularly: ammonia, nitrates

Substances Directive, and the grey list is List II of that directive. These directives are the 'parent' directives setting out the framework for the control of waste discharges in Europe, and are updated by the addition of 'daughter' directives as new compounds are identified as being harmful (e.g. PCP (pentachlorophenol)). The UK, along with other countries whose estuaries enter the North Sea, has in 1987 adopted a 'red' list (Table 6.4) of substances whose control is a priority, which will be controlled by cutbacks in all emissions irrespective of the water body to which they are discharged. The adoption of the 'red' list is a clear move towards a UES (or precautionary) policy for those substances which are regarded as the most hazardous to the marine and estuarine environment.

To control the discharge of material which may harm shellfish, environmental quality standards (EQSs) for waters with shellfish populations have been designated in both USA and Europe (Tables 6.5 and 6.6). Unlike the standards previously listed, these standards include bacterial or

Table 6.4 The red list. Substances whose control is a priority, which will be controlled in the North Sea, and its estuaries, by cutbacks in all emissions irrespective of the water body to which they are discharged. Source: North Sea Ministerial Conference, 1987.

Mercury
Cadmium
gamma—Hexachlorocyclohexane (Lindane)
DDT
Pentachlorophenol (PCP)
Hexachlorobenzene (HCB)
Hexachlorobutadiene (HCBD)
Aldrin
Dieldrin
Endrin
Chloroprene
3-Chlorotoluene
PCB (Polychlorinated biphenyls)
Triorganotin compounds
Dichlorvos
Trifluralin
Chloroform
Carbon tetrachloride
1, 2-Dichloroethane
Trichlorobenzene
Azinphos-methyl
Fenitrothion
Malathion
Endosulfan
Atrazine
Simazine

Table 6.5 Shellfish-rearing waters standards for the State of Virginia, USA. Source: Environmental Protection Agency (EPA) of the USA.

Parameter	
Dissolved oxygen	4–5 $mg\,l^{-1}$
pH	6–8.5
Total coliforms	70–100 ml^{-1}
Faecal coliforms	14–100 ml^{-1}
Mercury	0.1 $\mu g\,l^{-1}$
Cadmium	5.0 $\mu g\,l^{-1}$
DDT	1.0 $ng\,l^{-1}$
Dieldrin	3.0 $ng\,l^{-1}$

Table 6.6 Environmental quality standards for Europe on the quality of water required by designated Shellfish waters. Source: EC Directive (79/923), supplemented by guidelines adopted for Scotland. Mandatory means that the values given should not be exceeded.

Parameter	*Mandatory EQS*	*Notes*
pH	7–9	
Coloration	Deviation < 10 mg/Pt/1	
Suspended solids	Increase < 30% background	
Salinity	< 40‰ increase, < 10% background	
Dissolved oxygen	< 60%, average > 70% saturation	
Petroleum	No visible film	
Hydrocarbons	No harmful effects	
Organohalogens	No harmful effects*1	
Trace metals	No harmful effects*2	
Faecal coliforms	< 300/100 ml	Applied where live shellfish directly edible by man
Substances affecting taste	Below taste threshold in fish	Applied where presence is presumed

*1, *2 *Trace metal and organohalogenated substances: levels below which harmful effects will not occur*

Element	*Annual mean conc* $\mu g\ l^{-1}$	*Substance*	*Flesh conc ng/g wet wt where lipid conc* $\leqslant 1\%$
Ag	0.3	Dieldrin	50
As	20	DDD*	100
Cd	1.0	DDE*	100
Cr	10	DDT*	100
Cu	5.0	HCB	100
Hg	0.1	αHCH	30
Ni	5.0	γHCH	30
Pb	5.0	PCB	1000
Zn	10.0	Toxaphene	1000

*Sum of DDT 250 ng/g

Table 6.7 Bathing waters standards. Expressed as numbers per 100 ml. *Mandatory means that it should not be exceeded, Guideline indicates the desirable aims. Source: Environmental Protection Agency (EPA) of the USA, and European Community Directives.

Parameter	*USA*	*Europe Mandatory**	*Europe Guideline**
Total coliforms	700	10000	500
Faecal coliforms	200	2000	100
Faecal streptococci	—	300	100
Enteroviruses	—	0	0
Salmonella	—	0	0

microbial contamination. Standards, again based on bacterial or microbial contamination have also been adopted for beaches used for public bathing (Table 6.7).

6.3 Planning

When the policies of management in estuaries are implemented as planning objectives, the concepts of EQO, EQS and UES are perhaps not as different as they first appear. In practice they usually work together, in that the estuarine manager needs to specify the permitted standards in any effluent entering the estuary, in order to achieve the objectives which have been set for that estuary. Whether the standards set are uniform in all countries, or parts of that country, is essentially a political decision. The role of the estuarine scientist is to advise the politicians as to what standards are required, either in the effluent, or the receiving waters, so that the estuary is fit for its uses. In practice, in the context of estuarine water quality, the concentration of a particular constituent in a discharge may not be the most important factor; it is the rate of entry to the estuary that matters—this is obtained by multiplying the concentration by the rate of flow to give the 'load'. The assessment of loads is, however, often difficult since many of the sources of pollution are subject to wide variations in both flow and composition (Gameson, 1982). As the effluent load enters the estuary it is usual to define a 'zone of acceptable ecological impact', or a 'mixing zone', which is an area where dilution and dispersion occurs. The zone should be as small as practicable, and some latitude in the EQS for this area may often be deemed acceptable.

The usual method of implementing an estuarine management programme, is first to acquire knowledge of the state of the estuary, the quality of its waters, and the condition of its inhabitants at all times of the year. The discharges from industries, rivers or domestic sewage-works which enter into the estuary should be identified and analysed. A specification or 'consent' for each discharge should be issued by the relevant authority, to the discharger specifying the desirable composition of the effluent, and the maximum permitted volume or flow of discharge, hence controlling the load entering the estuary. Such consents should take account of international and national standards and agreements, such as whether the discharge contains any of the 'black', 'grey' or 'red' list substances, as well as the local situation of the discharge. In practice it may not be possible to have a complete knowledge of the estuary before issuing consent specifications, and the estuary will almost certainly not be pristine. In such cases it would be necessary to issue a provisional consent, for a fixed period of time, and to modify it in the light of subsequent knowledge about the condition of the estuary, or the effect of the effluent.

Toxicity testing (see Chapter 5) has a role for estuarine management in helping to determine the permitted levels of toxicant in the discharge. It has been considered in the UK that 1/10 of the 96h LC_{50} is a 'safe level' for the discharge of a toxic effluent, whilst in the US the use of 1/100 of the 96h LC_{50} has been advised. The factors 1/10 or 1/100 are termed 'application factors'. Such application factors have been criticised, because exposure to sub-lethal concentrations of a pollutant may influence physiological processes such as reproduction or development, but not cause death. However, despite such valid criticism, the acute toxicity test and application factors remain in use, since such tests are easy and reliable to perform, and provided that the most sensitive life stage of an animal is used, do give valuable guidelines for determining priorities in the management of industrial effluents discharged to estuarine ecosystems.

In the USA the Environmental Protection Agency (EPA), and the Coastal Zone Management Act (CZMA) have been the main federal instruments for the protection of coastline environments. Since 1972, all 35 coastal states have participated in the CZMA programme. Each individual state has produced legislation for its own shoreline, and inevitably it has proved hard to maintain Federal consistency, so that standards for estuarine management do vary from state to state.

Within Britain, a diversity of organisations are involved, and the British approach has been criticised as being uneven and often ill-defined (Carter, 1988). The Government Department of the Environment, and Scottish Development Department, oversee the local Water Authorities (soon to be replaced by a National Rivers Authority), or River Purification Boards, who are responsible for issuing consents for individual discharges. Dredging and the dumping of waste at sea are controlled by the Ministry of Agriculture, Fisheries and Food, and the Department of Agriculture and Fisheries for Scotland. The Nature Conservancy Council, along with bodies such as the Countryside Commission, represent the voice of nature conservation, by declaring estuarine sites as nature reserves, or sites of special scientific interest (SSIs), often because of the need to protect bird or fish populations and their food supply. In the Netherlands, a strong government policy on estuarine and coastal zone management is in operation, implemented by the Rijkswaterstaat, especially the Tidal Waters Division.

6.4 Practice

The management of estuaries has in some cases produced spectacular reversals in the conditions of estuaries after decades, or even centuries, of neglect and pollution. The cleanup of the Delaware Estuary represents one of the premier water pollution control success stories in the United States (Albert, 1988). The Delaware Estuary, bounded by the states of Delaware, New Jersey and Pennsylvania, is located in one of the most complex urban-industrial regions of the United States (Figure 6.1), having one of the world's greatest concentrations of heavy industry, and America's second busiest port, and second largest complex of oil-refining and petrochemical plants. A population of 5.7 million people reside in the region. Once largely devoid of aerobic aquatic life, now, thanks to a long-term improvement programme, it supports a variety of recreational uses, year-round fish populations and a number of key migratory fish species.

Improving the quality of the Delaware Estuary has been an evolutionary process spanning almost 200 years, which is summarised in Table 6.8. When the Delaware was discovered by Henry Hudson in 1609 it was presumably pristine, but by 1799, when the first pollution survey was undertaken, a variety of pollution sources were noted, and in response to pollution of drinking water supplies, the first municipal water supply and sewer systems were constructed. As population, and disease grew, improved water supply and sewage systems were installed, which solved public health concerns, but did nothing about the pollution of the estuary. The greatest degradation of the estuary occurred in the first half of the twentieth century, following expansion of industry and population. From 1914 to 1937 surveys indicated that the estuary was substantially polluted, with

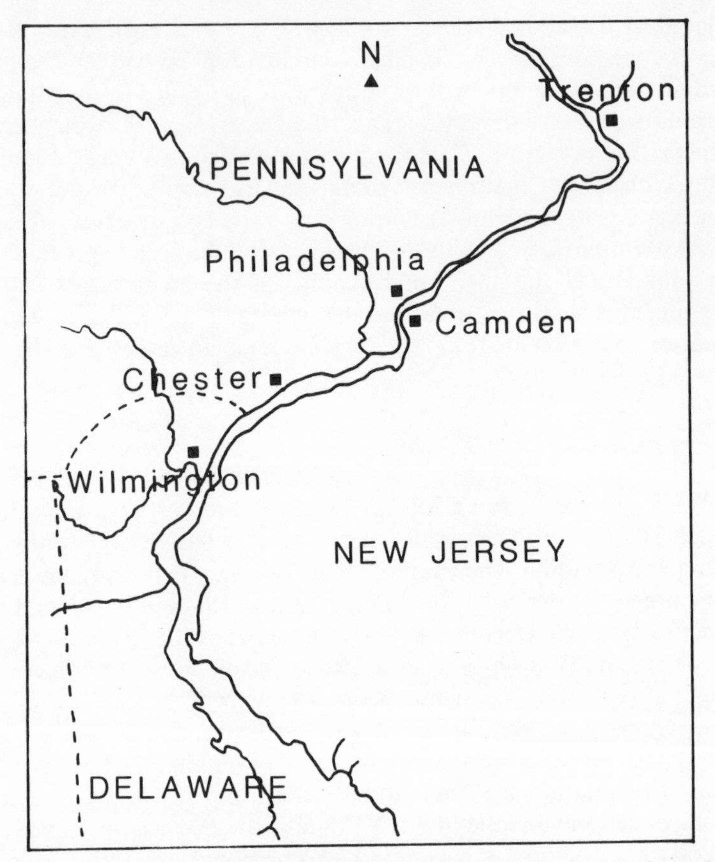

Figure 6.1 The Delaware estuary, eastern USA, showing the main centres of population. (After Albert, 1988.)

oxygen levels only just above zero. By 1946 an area of anoxia had developed running shore-to-shore and down the estuary for 20 miles (35 km) (Figure 6.2). Remedial action undertaken thereafter, involving the cleanup of rivers and streams entering the estuary, substantial reduction in organic discharges to the estuary and dredging of coal silt was undertaken, all with the aim of restoring dissolved oxygen levels to at least 50% saturation. As a result of these activities, the estuary became no longer anoxic, but it was still experiencing depressed dissolved oxygen levels, indicating continuing organic pollution (Figure 6.2). In the early 1960s new emission standards were adopted, which required the upgrading of most

Table 6.8 Description of the five generations of Delaware Estuary water pollution control efforts. From Albert (1988).

Generation	*Approximate time span*	*Problem*	*Actions*	*Prime participants**
First	1800–1860	Pollution of local water sources	Construction of municipal water systems with river intakes, some sewer line construction	Municipal government
Second	1880–1910	Water borne disease from consumption of river water	Construction of water filtration plants, development of alternative water supplies, sanitary sewer system construction	Municipal government
Third	1936–1960	Gross pollution	Construction of primary wastewater treatment plants after effluent standards adopted	INCODEL, states
Fourth	1960–1980	Substantial pollution	Construction of secondary or higher wastewater treatment plants after wasteloads allocated	DRBC, states and US EPA
Fifth	1980–?	Public health and aquatic life concerns including toxics	Underway. Could include combined sewer correction, more stringent point source controls, non-point source controls, toxic materials controls and other	DRBC, states and US EPA

*The efforts of the federal government prior to the creation of the United States Environmental Protection Agency (USEPA) and the efforts of most cities and industries to clean up their wastes in the third and fourth generation effort are recognised as well.
INCODEL = Interstate Commission on the Delaware River Basin.
DRBC = Delaware River Basin Commission.

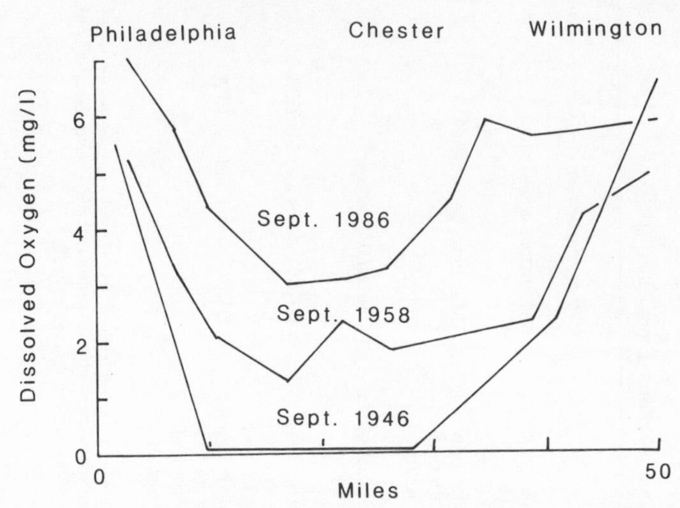

Figure 6.2 The dissolved oxygen concentration in the Delaware estuary, in 1946, 1958 and 1986. Measurements were made under similar conditions of tide, temperature and freshwater discharge. For locations see Figure 6.1. (After Albert, 1988.)

sewage-works from having primary treatment (i.e. removal of solids), to full secondary treatment (i.e. filtration and biological breakdown). As a result of these major works the dissolved oxygen concentration has risen steadily (Figures 6.2, 6.3), and the faecal coliform levels have dropped (Figure 6.4). There is still a seasonal sag in the oxygen content of the water (Figure 6.3), but as a result of these improvements the diversity of fish has increased from 16 resident species in the 1960s, to 36 by 1985. Improvements are still required, both to dissolved oxygen levels, and to the control of specific toxic substances, but the Delaware Estuary has been established as a prime example of how mankind can restore or repair the damage which he has caused to the estuarine ecosystem.

The Thames Estuary, UK, with the city of London and approximately 10 million people on its banks, has experienced similar problems to the Delaware, and like it has now been dramatically restored. On the Thames, there was a steady decline in the oxygenation of the estuary from 1920 until 1955, so that periods of total de-oxygenation regularly occurred. Since that date improvements, mainly to sewage-works, have resulted in a substantial improvement in the condition of the middle reaches. The recovery of the diversity of fish species in the Thames, as was shown in Figure 5.8, amply shows how successful has been the management programme here.

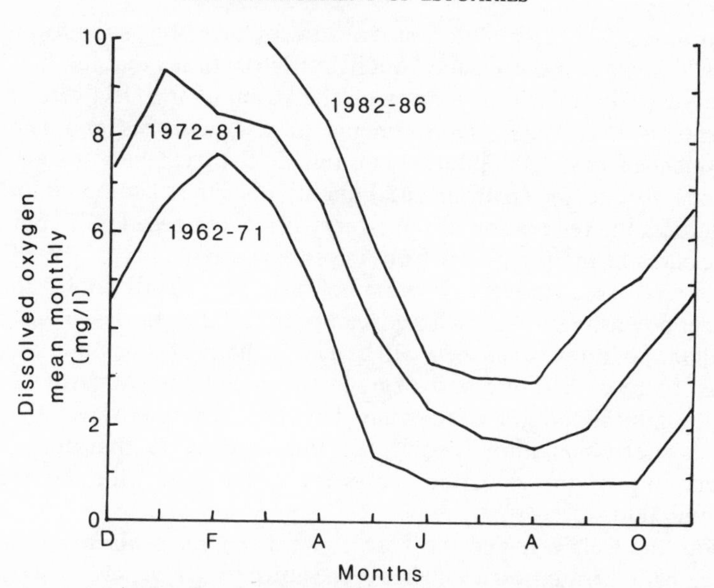

Figure 6.3 Mean monthly dissolved oxygen concentration in the Delaware estuary at Chester, averaged for each of three decades. Note the decrease in the duration of the critical low dissolved oxygen period with time. (After Albert, 1988.)

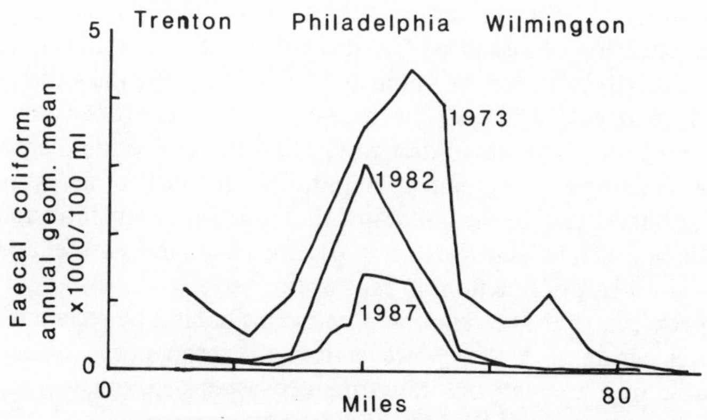

Figure 6.4 Annual geometric mean count of faecal coliform bacteria in the Delaware estuary for 1973, 1982 and 1987. (After Albert, 1988.)

Fortunately not all polluted estuaries are, or have been, as badly polluted as the Delaware or the Thames, but it is also true to say that few have been as successfully treated as those two. The lesson of the Delaware and the Thames is that major improvements to sewage-works and improved treatment of industrial effluents can restore the oxygenation state of an estuary, so that the environmental quality objectives can be attained. In both cases the restoration of the estuary has taken decades of effort, and large sums of money. Apart from improving the quality of the effluents discharged these recovery programmes have also clearly shown that the further downstream the discharges can be made, then the lesser will be the resultant pollution, since there will be a larger body of constantly changing water to receive, dilute and degrade the pollutants. In both of these estuaries the discharges to the estuary have been reduced to levels that were within the assimilation capacity of the estuary, so that the normal functioning of the estuarine ecosystem could cope with the material entering it.

Both the Delaware and the Thames were extreme problems of organic enrichment in estuaries, which has been shown to be reversible in its effects. Estuarine systems are generally resilient and will usually recover quite quickly once the source of organic pollution is controlled. However, the problems of toxic chemicals, especially those accumulative substances on the black lists, may be less readily reversible, and will continue to pose problems to estuarine managers. Many substances such as heavy metals will take 20–30 years to reduce their concentration to half of their concentration at the time of discharge (half-life). Even then the recovery of the ecosystem is largely dependent on the processes of sedimentation. Thus the concentration of the pollutant in the water column may decline rapidly after the cessation of discharge, but its concentration within the sediment may continue to influence the fauna and flora inhabiting the sediment, and a sediment-based EQS may be necessary. Any re-suspension of the sediment, due for example to dredging, will serve to liberate the pollutant into the water again. The intertidal and other habitats of an estuary can utilise (or bury) many of the pollutants that human activity introduces into estuaries, but such utilisation (or burial) depends on the maintenance of all aspects of a healthy functioning ecosystem.

The problems of the destruction of estuarine habitat by reclamation and other engineering works may however, in the long run, prove to be the most intractable in managing our estuarine ecosystems. Management has to include an assessment of the impact of loss of area, not only as a food loss for birds or fish, but also for the loss of a 'free sewage-works' area which can cope with the various pollutants discussed above. It is essential to understand the functioning of all aspects of the area before undertaking any irreversible activities.

The conservation interest in estuaries is focused on two aspects. Firstly the maintenance of fisheries, either by the harvesting of fish and shellfish, or by the use of estuaries as nursery areas for marine fish stocks, or for fish such as sprat and flounder caught within the estuary, or for estuaries as routes for migratory species to pass through. The conservation of fish populations is inhibited by poor water quality (usually low dissolved oxygen, but also for example high ammonia levels) which acts as a barrier to migratory fish, as well as levels of persistent pollutants, and high nutrient levels, which after the collapse of a phytoplankton bloom can lead to low dissolved oxygen levels. Secondly, the conservation interest is centred on the abundant bird populations which feed on the intertidal areas, attracted by the rich and available food supplies. To protect these interests, many estuaries have been declared as nature reserves, primarily because, as stated in the original paradoxes, they can be some of the least disturbed natural areas with rich natural animal populations, and also to help protect them against pollution or reclamation which can destroy them forever. The conservation of wildlife in estuaries need not exclude other users. Industries are compatable with wildlife, provided that organic effluents and power-station discharges are well diluted and carefully sited. Indeed small amounts of organic effluents, and well controlled power-station discharges may actually serve to increase the productivity of an estuary, and enhance its conservation interest. Persistent chemicals which accumulate in the food chain must however be strictly controlled. Excessive reclamation, especially if it takes away intertidal areas, which are often more productive than subtidal areas, poses probably the greatest single threat to estuarine conservation, destroying forever the habitat.

The management of estuaries is thus largely the protection of biological systems, even if the biology is that of mankind. The future management of estuaries as a resource for wildlife as well as for mankind depends very much on the successful implementation of legislation for pollution control and habitat protection.

READING LIST

The reading list for this book consists of first a selection of general reference books, multi-author works and journals which have applicability to all the book, or to several chapters, followed by references to individual papers which are arranged in chapter groupings. If a reference relates to more than one chapter it is listed in the first chapter that it appears in.

Books

BOADEN, P.J.S. and SEED, R. (1985) *An introduction to coastal ecology* Blackie, Glasgow, 224 pp.

DYER, K.R. (1973) *Estuaries: A physical introduction* John Wiley, London, 140 pp.

DYER, K.R. (1986) *Coastal and estuarine sediment dynamics* Wiley, Chichester, 342 pp.

GRAY, J.S. (1981) *The ecology of marine sediments* Cambridge University Press, Cambridge, 185 pp.

KNOX, G.A. (1986) *Estuarine ecosystems* (2 volumes) CRC Press, Florida, 289 + 230 pp.

MANN, K.H. (1982) *Ecology of Coastal Waters; A Systems Approach* Blackwell Sci. Pub., Oxford, 322 pp.

MEADOWS, P.S. and CAMPBELL, J.I. (1987) *An Introduction to Marine Science* (2nd Edition) Blackie, Glasgow, 288 pp.

MUUS, B.J. (1967) The fauna of Danish estuaries and lagoons. Distribution and ecology of dominating species in the shallow reaches of the mesohaline zone. *Medd. fra. Danmarks fiskeri og Havundersogelser. Ny serie*, 5:1 315 pp.

PERKINS, E.J. (1974), *The Biology of Estuarine and Coastal Waters* Academic Press, London, 678 pp.

REISE, K. (1985) *Tidal flat ecology—an experimental approach* Springer-Verlag, Heidelberg, 191 pp.

WOLFF, W.J. (1973) The estuary as a habitat. An analysis of data on the soft-bottom macrofauna of the estuarine area of the Rivers Rhine, Meuse and Scheldt. *Zoologische Verhandelingen*, **126**, 242 pp.

Multi-author works

BURTON, J.D. and LISS, P.S. (editors) (1976) *Estuarine chemistry* Academic Press, London, 229 pp.

COULL, B.C. (editor) (1977) *Ecology of marine benthos* Belle W. Baruch Library in Marine Science. No. 6, Univ. of S. Carolina Press, 467 pp.

CHAPMAN, V.J. (editor) (1977) *Wet coastal ecosystems* Elsevier, Amsterdam, 428 pp.

DAY, J.H. (editor) (1980) *Estuarine ecology, with particular reference to southern Africa* Balkema books, Rotterdam, 400 pp.

JEFFERIES, R.L. and DAVY, A.J. (editors) (1979) *Ecological processes in coastal environments* Blackwell Sci. Pub., Oxford, 684 pp.

JONES, N.V. and WOLFF, W.J. (Editors) (1981) *Feeding and survival strategies in estuarine organisms* Plenum Press, New York, 304 pp.

KENNEDY, V.S. (editor) (1980) *Estuarine perspectives* Academic Press, New York, 533 pp.
KENNEDY, V.S. (editor) (1982) *Estuarine comparisons* Academic Press, New York, 709 pp.
KENNEDY, V.S. (editor) (1984) *The estuary as a filter* Academic Press, New York.
KETCHUM, B.H. (editor) (1983) *Estuaries and Enclosed Seas* Elsevier, Amsterdam, 500 pp.
KINNE, O. (editor) (1978, onwards) *Marine Ecology* 5 volumes, John Wiley, New York.
KJERFVE, B. (editor) (1978) *Estuarine transport processes* Belle W. Baruch Library in Marine Science, No. 7, University of S. Carolina Press, 331 pp.
KNOX, G.A. and KILNER, A.R. (1973) *The ecology of the Avon-Heathcote estuary* University of Canterbury, New Zealand, 358 pp.
LAUFF, G.H. (editor) (1967) *Estuaries* American Association for the Advancement of Science, Publication No. 83, 755 pp.
LIPPSON, A.J. *et al.* (1981) *Environmental atlas of the Potomac estuary* Available from Johns Hopkins University Press, Maryland, 280 pp.
POSTMA, H. and ZIJLSTRA, J.J. (editors) (1988) *Continental shelves* Elsevier, Amsterdam, 421 pp.
RIJKSWATERSTAAT (1985) *Biological research in the ems-dollart estuary* Rijkswaterstaat, The Hague, Netherlands, 182 pp.
TENORE, K.R. and COULL, B.C. (1980) (editors) *Marine benthic dynamics* Belle W. Baruch Library in Marine Science, No. 11, Univ. of S. Carolina Press, 451 pp.
WILEY, M. (editor) (1976) *Estuarine processes* 2 volumes, Academic Press, London, 603 pp.
WOLFE, D.A. (editor) (1986) *Estuarine variability* Academic Press, Florida, 509 pp.
WOLFF, W.J. (editor) (1980) *Ecology of the Wadden Sea* Balkema Books, Rotterdam, 1300 pp in 3 volumes.

Handbooks on methods for studying estuaries

BAKER, J.M. and WOLFF, W.J. (editors) (1987) *Biological surveys of estuaries and coasts* EBSA handbook, Cambridge University Press, Cambridge, 449 pp.
DYER, K.R. (editor) (1979) *Estuarine hydrography and sedimentation* EBSA handbook, Cambridge University Press, Cambridge, 230 pp.
HEAD, P.C. (editor) (1985) *Practical estuarine chemistry* EBSA handbook, Cambridge University Press, Cambridge, 337 pp.
HOLME, N.A. and McINTYRE, A.D. (1984) *Methods for the study of marine benthos (second edition)* Blackwell Sci. Pub., Oxford, 387 pp.
LIPPSON, A.J. and R.L. (1984) *Life in the Chesapeake Bay* John Hopkins University Press, Baltimore, Maryland, 230 pp.
MORRIS, A.W. (editor) (1983) *Practical procedures for estuarine studies* Natural Environment Research Council, Plymouth, 262 pp.

Journals

The publication of material relating to the Estuarine Ecosystem has been greatly stimulated by the Estuarine and Coastal Sciences Association (mainly for UK and Europe) and the Estuarine Research Federation (mainly for USA) who organise regular meetings and publish bulletins and the journals *Estuarine and Coastal Marine Science* and *Estuaries*. Any reader who would like details of these associations is invited to write to the author who will forward the current address of the ECSA or ERF Secretary.

In addition to the above journals, the following journals contain a selection of relevant papers.

Aquatic Science and Biology Abstracts
Estuaries and coastal waters of the British Isles; Annual bibliography, M.B.A. Plymouth

Journal of Animal Ecology
Journal of Applied Ecology
Journal of Experimental Marine Biology and Ecology
Journal of the Marine Biological Association of the U.K.
Lecture notes on Coastal and Estuarine Studies (Springer-Verlag)
Limnology and Oceanography
Marine Biology
Marine Ecology, Progress Series
Marine Pollution Bulletin
Netherlands Journal of Sea Research
Oceanography and Marine Biology: An Annual Review
Ophelia

References

Chapter One

ALEXANDER, W.B., SOUTHGATE, B.A. and BASSINDALE, R. (1935) Survey of the River Tees. Part II—the estuary, chemical and biological *D.S.I.R., Water Pollution Research Technical Paper*, **5**, 1–171.

ANDERSON, F.E. (1983) The nothern muddy intertidal: a seasonally changing source of suspended sediments to estuarine waters—a review *Canadian Journal of Fisheries and Aquatic Sciences*, **40** (Supp 1), 143–159.

ANSELL, A.D., McLUSKY, D.S., STIRLING, A. and TREVALLION, A. (1978) Production and energy flow in the macrobenthos of two sandy beaches in S.W. India *Proceedings of the Royal Society of Edinburgh*, **76B**, 269–296.

BALE, A.J. and MORRIS, A.W. (1987) *In situ* measurement of particle size in estuarine waters *Estuarine, Coastal and Shelf Science*, **24**, 253–263.

BREWERS, J.M. and YEATS, P.A. (1978) Trace metals in the waters of a partially mixed estuary *Estuarine and Coastal Marine Science*, **7**, 147–162.

BREY, T., RUMOHR, H. and ANKAR, S. (1988) Energy content of macrobenthic invertebrates: general conversion factors from weight to energy *Journal of experimental marine biology and ecology*, **117**, 271–278.

BIGGS, R.B. (1970) Sources and distribution of suspended sediment in North Chesapeake Bay *Marine Geology*, **9**, 187–201.

BOKUNIEWICZ, H.J., GEBERT, J. and GORDON, R.B. (1976) Sediment mass balance of a large estuary, Lond Island Sound *Estuarine and Coastal Marine Science*, **4**, 523–536.

BUCHANAN, J.B. and WARWICK, R.M. (1974) An estimate of benthic macrofaunal production in the offshore mud of the Northumberland Coast *Journal of the Marine Biological Association of the UK.*, **54**, 197–222.

CARRIKER, M.R. (1967) Ecology of estuarine benthic invertebrates: a perspective, in Lauff (ed.) *Estuaries*. AAAS, **83**, 442–487.

CHAMBERS, M.R. and MILNE, H. (1979) Seasonal variation in the condition of some intertidal invertebrates of the Ythan estuary *Estuarine and Coastal Marine Science*, **8**, 411–419.

CRISP, D. (1971) Energy flow measurements, in Holme, N.A and McIntyre, A.D. (eds) *Methods for Studying the Marine Benthos*, IBP handbook 16, Blackwell, Oxford, 197–280.

DEATON, L.E. and GREENBERG, M.J. (1986) There is no horohalinicum *Estuaries*, **9**, 20–30.

ELLIOTT, A.J. (1978) Observations of the meterologically induced circulation in the Potomac estuary *Estuarine and Coastal Marine Science*, **6**, 285–299.

FESTA, J.F. and HANSEN, D.V. (1978) Turbidity maxima in partially mixed estuaries: A two-dimensional numerical model *Estuarine and Coastal Marine Science*, **7**, 347–359.

GORSLINE, D.S. (1967) Contrasts in coastal bay sediments on the Gulf and Pacific coasts, in Lauff (ed) *Estuaries*, AAAS, **83**, 219–225.

HARRISON, S.J. and PHIZACKLEA, A.P. (1987) Vertical temperature gradients in muddy intertidal sediments in the Forth estuary, Scotland *Limnology and Oceanography*, **32**, 954–963.

HARTWIG, E.O. (1976) Nutrient cycling in marine sediments I Organic carbon *Marine Biology*, **34**, 285–295.

HEAD, P.C. (1976) Organic processes in estuaries, in Burton J.D. and Liss, P.S. (ed.) *Estuarine chemistry*, Academic Press, London, 54–91.

HEDGPETH, J.W. (1967) The sense of the meeting, in Lauff (ed.) *Estuaries*, AAAS, **83**, 707–712.

HUGHES, R.N. (1970) An energy budget for a tidal flat population of the bivalve *Scrobicularia plana Journal of Animal Ecology*, **39**, 357–381.

KENNISH, J.R. (1986) *Ecology of estuaries (Volume 1, Physical and Chemical aspects)*, CRC Press, Florida, 254 pp.

KHAYRALLA, N. and JONES, A.M. (1975) A survey of the benthos of the Tay estuary *Proceedings of the Royal Society of Edinburgh*, **75B**, 113–135.

KHLEBOVICH, V.V. (1968) Some peculiar features of the hydrochemical regime and the fauna of mesohaline waters *Marine Biology*, **2**, 47–49.

KING, C.M. (1975) *Introduction to Marine Geology and Geomorphology*, E. Arnold, London, 370 pp.

KRONE, R.B. (1978) Aggregation of suspended particles in estuaries, in Kjerfve, B. (ed) *Estuarine transport processes*, Univ. S. Carolina Press, 177–190.

JARVIS, J. and RILEY, C. (1987) Sediment transport in the mouth of the Eden estuary *Estuarine, Coastal and Shelf Science*, **24**, 463–481.

LEACH, J.H. (1971) Hydrology of the Ythan estuary with reference to distribution of major nutrients and detritus *Journal of the Marine Biological Association of the UK.*, **51**, 137–157.

LINDEMANN, R.L. (1942) The trophic-dynamic aspect of ecology *Ecology*, **23**, 399–418.

MAITLAND, P.S. and HUDSPITH, P.M.G. (1984) The zoobenthos of Loch Leven, Kinross and estimates of its production in the sandy littoral area during 1970–71 *Proceedings of the Royal Society of Edinburgh*, **74B**, 219–240.

MANN, K.H. (1972) Macrophyte production and detritus food chains in coastal waters *Memorie dell' Instituto Italiano di Idrobiologia*, **29**, Suppl, 353–383.

MANTOURA, R.F.C. (1987) Organic films at the halocline *Nature*, **328**, 579–580.

McINTYRE, A.D. (1970) The range of biomass in intertidal sand, with special reference to *Tellina tenuis Journal of the Marine Biological Association of the UK.*, **50**, 561–576.

McLUSKY, D.S. (1987) Intertidal habitats and benthic macrofauna of the Forth estuary *Proceedings of the Royal Society of Edinburgh*, **93B**, 389–400.

McLUSKY, D.S. and HUNTER, R. (1985) Loch Riddon revisited—the intertidal of the sea-loch resurveyed after 53 years *Glasgow Naturalist*, **21**, 53–62.

McLUSKY, D.S., TEARE, M. and PHIZACKLEA, P. (1980) Effects of domestic and industrial pollution on the distribution and abundance of aquatic oligochaetes in the Forth estuary *Helgolander wiss. Meeresunters.*, **33**, 113–121.

McNEILL, S. and LAWTON, J.H. (1970) Annual production and respiration in animal populations *Nature*, **225**, 472–474.

METTAM, C. (1983) An estuarine mudflat resurveyed after 45 years *Oceanologica Acta, Proc 17th EMBS*, 137–140.

MORRIS, A.W. *et al.* (1978) Very low salinity regions of estuaries: important sites for chemical and biological reactions *Nature*, **274**, 678–680.

MUUS, B. (1974) Ecophysiological problems of the brackish water *Hydrobiological Bulletin*, **8**, 76–89.

PEARSON, T. and STANLEY, S.O. (1979) Comparative measurement of the redox potential of marine sediments as a rapid means of assessing the effect of organic pollution *Marine Biology*, **53**, 371–379.

PHILLIPS, J. (1972) Chemical processes in estuaries, in Barnes, R.S.K. and Green, J. (eds) *The estuarine environment*, Applied Science Publishers, London, 33–50.

PHILLIPSON, J. (1966) *Ecological energetics*, E. Arnold, London, 57 pp.

POPHAM, E.J. (1966) The littoral fauna of the Ribble estuary, Lancashire, England *Oikos*, **17**, 19–32.

POSTMA, H. (1967) Sediment transport and sedimentation in the estuarine environment, in Lauff (ed.) *Estuaries*, AAAS, **83**, 158–179.

PRITCHARD, D.W. (1967) What is an estuary: a physical viewpoint, in Lauff, G.H. (ed.) *Estuaries*, American Association for the Advancement of Science, Publication No. 83, 3–5.

QASIM, S.Z. and SEN GUPTA, R. (1981) Environmental characteristics of the Mandovi-Zuari estuarine system in Goa *Estuarine, Coastal and Shelf Science*, **13**, 557–578.

RANKIN, J.C. and DAVENPORT, J.A. (1981) *Animal osmoregulation*, Blackie, Glasgow, 220 pp.

REMANE, A. and SCHLIEPER, C. (1958) *Die biologie des brackwassers*, E. Schwiezerbart'sche verlagsbuchhandlung, Stuttgart, 348 pp.

RHOADS, D.C. (1974) Organism-sediment relations on the muddy sea floor *Oceanography and Marine Biology, Annual Review*, **12**, 263–300.

ROBERTS, W.P. and PIERCE, J.W. (1976) Deposition in the upper Patuxent estuary, Maryland, 1968–1969 *Estuarine and Coastal Marine Science*, **4**, 267–280.

SALONEN, K., SARVALA, J., HAKALA, I. and VILJANEN, M.-J. (1976) The relation of energy and organic carbon in aquatic invertebrates *Limnology and Oceanography*, **21**, 724–730.

SANDERS, H.L. (1969) Benthic marine diversity and the stability-time hypothesis *Brookhaven Symposia in Biology*, **22**, 71–81.

SCHWINGHAMER, P. *et al.* (1986) Partitioning of production and respiration among size groups of organisms in an intertidal benthic community *Marine Ecology Progress Series*, **31**, 131–142.

SHOLKOVITZ, E.R. (1979) Chemical and physical processes controlling the chemical composition of suspended material in the River Tay estuary *Estuarine and Coastal Marine Science*, **8**, 523–545.

de SILVA SAMARASINGHE, J.R. and LENNON, G.W. (1987) Hypersalinity, flushing and transient salt-wedges in a tidal gulf—an inverse estuary *Estuarine, Coastal and Shelf Science*, **24**, 483–498.

THIA-ENG, C. (1973) An ecological study of the Ponggal estuary, Singapore *Hydrobiologia*, **43**, 505–533.

VAN ES, F.B. (1977) A preliminary carbon budget for a part of the Ems estuary: The Dollard *Helgolander wiss. Meeresunters.*, **30**, 283–291.

VENICE SYSTEM (1959) Symposium on the classification of brackish waters, Venice, April 8–14, 1958. *Arch. Oceanog. Limnol.*, **11** (supplement), 1–248.

WATERS, T.F. (1969) The turnover ratio in production ecology of freshwater invertebrates *American Naturalist*, **103**, 173–185.

WILDISH, D.J. (1977) Factors controlling marine and estuarine sublittoral macrofauna *Helgolander wiss. Meeresunters.*, **30**, 445–454.

WINBERG, G.C. (1971) *Methods for the estimation of the production of aquatic animals*, Academic Press, London, 175 pp.

WOLFF, W.J. (1972) Origin and history of the brackish-water fauna of N.W. Europe *Proc. 5th Europ. Mar. Biol. Symp*, Piccin Editore, Padova, 11–18.

ZUTIC, V. and LEGOVIC, C. (1987) A film of organic matter at the fresh-water/sea-water interface of an estuary *Nature*, **328**, 612–614.

Chapter Two

BAILLIE, P.W. (1986) Oxygenation of intertidal estuarine sediments by benthic microalgal photosynthesis *Estuarine, Coastal and Shelf Science*, **22**, 143–159.

BOYNTON, W.R., KEMP, W.M. and KEEFE, C.W. (1982) A comparative analysis of nutrients and other factors influencing estuarine phytoplankton production, in Kennedy, V.S. (ed.) *Estuarine Comparisons*, Academic Press, New York, 69–90.

CADEE, G.C. (1971) Primary production on a tidal flat *Netherlands Journal of Zoology*, **21**, 213.

CHRISTOFI, N., OWENS, N.J.P. and STEWART, W.D.P. (1979) Studies on nitrifying micro-organisms of the Eden estuary, Scotland, in Naylor E. and Hartnoll, R. (eds) *Proceedings 13th Europ Mar. Biol. Symp.*, Pergamon Press, Oxford, 259–266.

CLOERN, J.E. (1982) Does the benthos control phytoplankton biomass in south San Francisco Bay? *Marine Ecology, Progress Series*, **9**, 191–202.

COLIJN, F. and DE JONGE, V.N. (1984) Primary production of microphytobenthos in the Ems-Dollard estuary *Marine Ecology, Progress Series*, **14**, 185–196.

DAME, R.F. and STILWELL, D. (1984) Environmental factors influencing macrodetritus flux in North Inlet estuary *Estuarine, Coastal and Shelf Science*, **18**, 721–726.

DARNELL, R.M. (1967) The organic detritus problem, in Lauff, G.H. (ed.) *Estuaries*, American Association for the Advancement of Science, Publication No. 83, 374–375.

DAWES, C.J., MOON, R.E. and Davies, M.A. (1978) The photosynthetic and respiratory rates and tolerance of benthic algae from a mangrove and salt marsh estuary *Estuarine and Coastal Marine Science*, **6**, 175–185.

EILERS, H.P. (1979) Production ecology in an Oregon coastal salt marsh *Estuarine and Coastal Marine Science*, **8**, 399–410.

ERIKSSON, S., SELLEI, P. and WALLSTROM, K. (1977) The structure of the plankton community of the Oregrundsgrepen (S.W. Bothian Sea) *Helgolander wiss. Meeresunters*, **30**, 582–597.

FENCHEL, T. (1972) Aspects of decomposer food chains in marine benthos *Sonderdruck aus Verhandlungsbericht der Deutschem Zoologischen Gesellschaft, 65 Jahresversammlung*, 14–21.

GALLACHER, J.L., PFEIFFER, W.J. and POMERO, L.R. (1976) Leaching and microbial utilisation of dissolved organic carbon from leaves of *Spartina alterniflora Estuarine and Coastal Marine Science*, **4**, 467–471.

GREENWAY, M. (1976) The grazing of *Thalassia testudinum* in Kingston Harbour, Jamaica *Aquatic Botany*, **2**, 117–126.

GUTERSTAM, B. (1977) An *in situ* study of the primary production and the metabolism of a Baltic *F. vesiculosus* community, in Keegan, B. *et al.* (eds) *Proceedings of 11th Europ. Mar. Biol. Symp.*, Pergamon Press, Oxford, 311–319.

HAPP, G., GOSSELINK, J.G. and DAVY, J.W. (1977) The seasonal distribution of organic carbon in a Louisiana estuary *Estuarine and Coastal Marine Science*, **5**, 695–705.

HARRISON, P.G. (1987) Natural expansion and experimental manipulation of seagrass (*Zostera* spp.) abundance and the response of infaunal invertebrates *Estuarine, Coastal and Shelf Science*, **24**, 799–812.

HEINLE, D.R. and FLEMER, D.A. (1976) Flows of materials between poorly flooded tidal marshes and an estuary *Marine Biology*, **35**, 359–373.

JACKSON, R.H., WILLIAMS, P.J. le B. and JOINT, I.R. (1987) Freshwater phytoplankton in the low salinity region of the River Tamar estuary *Estuarine, Coastal and Shelf Sicience*, **25**, 299–311.

JEFFERIES, R.L. (1972) Aspects of salt-marsh ecology with particular reference to inorganic plant nutrition, in Barnes, R.S.K. and Green, J. (eds) *The Estuarine Environment*, Applied Science Publishers, London, 61–85.

JOINT, I.R. (1978) Microbial production of an estuarine mudflat *Estuarine and Coastal Marine Science*, **7**, 185–195.

JOREGENSEN, N.O.G., MOPPER, K. and LINDROTH, P. (1980) Occurrence, origin and assimilation of free amino acids in an estuarine environment *Ophelia*, Suppl 1, 179–192.

KIRBY-SMITH, W.W. (1976) The detritus problem and the feeding and digestion of an estuarine organism, in Wiley, M. (ed.), *Estuarine processes (I)*, Academic Press, London, 469–479.

KEEFE, C.W. (1972) Marsh production: a summary of the literature *Contributions in Marine Science, Texas*, **16**, 163–181.

KHAFJI, A.K. and NORTON, T.A. (1979) The effects of salinity on the distribution of *Fucus ceranoides Estuarine and Coastal Marine Science*, **8**, 433–439.

LAANE, R.W.P.M. (1982) Chemical characteristics of the organic matter in the waterphase of the Ems-Dollard estuary *Biologisch onderzoek eems-dollard estuarium*, **6**, 1–134.

LAPPALAINEN, A. (1973) Biotic fluctuations in a *Zostera marina* community *Oikos*, **15**, 74–80.

LIVELY, J.S., KAUFMAN, Z. and CARPENTER, E.J. (1983) Phytoplankton ecology of a barrier island estuary: Great South Bay, New York *Estuarine, Coastal and Shelf Science*, **16**, 51–68.

LONG, S.P. and MASON, C.F. (1983) *Saltmarsh ecology*, Blackie, Glasgow, 168 pp.

MALONE, T.C. (1977) Environmental regulation of phytoplankton productivity in the Lower Hudson estuary *Estuarine and Coastal Marine Science*, **5**, 157–171.

McROY, C.P. and HELFFERICH, C. (editors) (1977) *Seagrass ecosystems: A scientific perspective*, Marcel Dekker, New York, 314 pp.

MELCHIORI-SANTOLINI, U. and HOPTON, J.W. (editors) (1972) *Detritus and its role in aquatic ecosystems*, Memorie dell' Instituto Italiano di Idrobiologia, 29 (suppl), 540 pp.

ODUM, E.P. and DE LA CRUZ, A.A. (1967) Particulate organic detritus in a Georgia salt marsh/estuarine ecosystem, in Lauff (ed.) *Estuaries*, AAAS, **83**, 383–394.

PICKVAL, J.C. and ODUM, W.E. (1977) Benthic detritus in a salt marsh tidal creek, in Wiley, M. (ed.) *Estuarine processes (II)*, Academic Press, London, 280–292.

PIERCE, S.M. (1983) Estimation of the non-seasonal production of *Spartina maritima* in a South African estuary *Estuarine, Coastal and Shelf Science*, **16**, 241–254.

POMEROY, L.R. *et al.* (1977) Flow of organic matter through a salt marsh, in Wiley, M. (ed.) *Estuarine processes (II)*, Academic Press, London, 270–279.

POSTMA, H. (1988) Tidal flat areas, in Jansson, B-O. (ed.) *Coastal-Offshore Ecosystem Interactions*, Springer-Verlag, Berlin, 102–121.

RASMUSSEN, E. (1973) Systematics and ecology of the Isefjord marine fauna *Ophelia*, **11**, 1–495.

ROBERTSON, A.I. (1988) Decomposition of mangrove leaf litter in tropical Australia *Journal of experimental marine biology and ecology*, **116**, 235–247.

SAND-JENSEN, K. (1975) Biomass, net production and growth dynamics in an eelgrass (*Zostera marina*) population in Vellerup Vej, Denmark *Ophelia*, **14**, 185–201.

STEVENSON, J.C. *et al.* (1977) Nutrient exchanges between brackish water marshes and the estuary, in Wiley, M. (ed.) *Estuarine Processes (II)*, Academic Press, London, 219–240.

STRACHAL, G. and GANNING, B. (1977) *Boken om havet*, Forskning och Framsteg, Stockholm, 132 pp.

TEAL, J.M. (1962) Energy flow in the salt marsh ecosystem of Georgia *Ecology*, **43**, 614–62.

TENORE, K.R. (1981) Organic nitrogen and caloric content of detritus (I) *Estuarine, Coastal and Shelf Science*, **12**, 39–47.

VALIELA, I. and TEAL, J.M. (1979) The nitrogen budget of a salt marsh ecosystem *Nature*, **280**, 652–656.

VAN RAALTE, C.D., VALIELA, I. and TEAL, J.M. (1976) Production of epibenthic salt marsh algae: light and nutrient limitation *Limnology and Oceanography*, **21**, 862–872.

VAN VALKENBERG, S.D. *et al.* (1978) A comparison by size class and volume of detritus versus phytoplankton in Chesapeake Bay *Estuarine and Coastal Marine Science*, **6**, 569–582.

WILLIAMS, P.J. le B. (1981) Primary productivity and heterotrophic activity in estuaries, in *River Inputs to Ocean Systems*, United Nations, New York, 243–258.

WILLIAMS, R.B. (1972) Annual phytoplanktonic production in a system of shallow temperate estuaries, in Barnes, H. (ed.) *Some contemporary studies in marine science*, Aberdeen Univ. Press, 699–716.

WOLFF, W.J. (1977) A benthic food budget for the Grevelingen estuary, The Netherlands, and a consideration of the mechanisms causing high benthic secondary production in estuaries, in Coull, B.C. (ed.) *Ecology of marine benthos*, Univ. S. Carolina Press, 267–280.

WOLFF, W.J. (1980) Biotic aspects of the chemistry of estuaries, in Olausson, E. and Cato, I. (eds) *Chemistry and biogeochemistry of estuaries*, John Wiley, New York, 264–295.

Chapter Three

ANDERSON, S.S. (1972) The ecology of Morecambe Bay (II) *Journal of Applied Ecology*, **9**, 161–177.

ANKAR, S. and ELMGREN, R. (1976) The benthic macro- and meio-fauna of the Askö-Landsort Area (Northern Baltic proper). A stratified random sampling survey *Contributions from the Askö Laboratory, Univ. Stockholm*, **11**, 1–115.

BARLOCHER, F., GORDON, J. and IRELAND, R.J. (1988) Organic composition of seafoam and its digestion by *Corophium volutator Journal of experimental marine biology and ecology*, **115**, 179–186.

BEUKEMA, J.J. (1976) Biomass and species richness of the macro-benthic animals living on the tidal flats of the Dutch Wadden Sea *Netherlands Journal of Sea Research*, **10**, 236–261.

BEUKEMA, J.J., CADEE, G.C. and HUMMEL, H. (1983) Differential variability in time and space of numbers in suspension feeding and deposit feeding benthic species in a tidal flat area *Oceanologica Acta*, special volume, 21–26.

BEUKEMA, J.J., KNOL, E. and CADEE, G.C. (1985) Effects of temperature on the length of the annual growing season in the tellinid bivalve *Macoma balthica* living on tidal flats in the Dutch Wadden Sea *Journal of experimental marine biology and ecology*, **90**, 129–144.

BERNARD, F.R. (1974) Annual biodeposition and gross energy budget of mature Pacific oysters, *Crassostrea gigas Journal of the Fisheries Research Board of Canada*, **31**, 185–190.

BLOOM, S.A., SIMON, J.L. and HUNTER, V.D. (1972) Animal-sediment relations and community analysis of a Florida estuary *Marine Biology*, **13**, 43–56.

BUHR, K.J. and WINTER, J.E. (1977) Distribution and maintenance of a *Lanice conchilega* association in the Weser estuary, with special reference to the suspension-feeding behaviour of *Lanice conchilega*, in Keegan, B. *et al.* (eds) *Proceedings of the 11th Europ. Mar. Biol. Symp.*, Pergamon Press, Oxford, 101–113.

BURKE, M.V. and MANN, K.H. (1974) Productivity and production biomass ratios of bivalve and gastropod populations in an eastern Canadian estuary *Journal of the Fisheries Research Board of Canada*, **31**, 167–177.

CHAMBERS, M.R. and MILNE, H. (1975) The production of *Macoma balthica* in the Ythan estuary *Estuarine and Coastal Marine Science*, **3**, 443–455.

CHAMBERS, M.R. and MILNE, H. (1975) Life cycle and production of *Nereis diversicolor* in the Ythan estuary, Scotland *Estuarine and Coastal Marine Science*, **3**, 133–144.

CRANFORD, P.J., PEER, D.L. and GORDON, D.C. (1985) Population dynamics and production of *Macoma balthica* in Cumberland basin and Shepody Bay, Bay of Fundy *Netherlands Journal of Sea Research*, **19**, 135–146.

DAME, R.F. (1976) Energy flow in an intertidal oyster population *Estuarine and Coastal Marine Science*, **4**, 243–253.

DE WILDE, P.A.W.J. (1975) Influence of temperature on behaviour, energy metabolism and growth of *Macoma balthica*, in Barnes, H. (ed.) *Proceedings of 9th Europ. Mar. Biol. Symp.*, Aberdeen Univ. Press, 239–259.

ELMGREN, R. (1984) Trophic dynamics in the enclosed, brackish Baltic Sea *Rapp. P.-v. Reun. Cons. int. Explor. Mer.*, **183**, 152–169.

EMERSON, C.W., MINCHINTON, T.E. and GRANT, T. (1988) Population structure, biomass and respiration of *Mya arenaria* on temperate sandflat *Journal of experimental marine biology and ecology*, **115**, 99–111.

FENCHEL, T., KOFOED, L.H. and LAPPALAINEN, A. (1975) Particle size-selection of two deposit feeders: the amphipod *Corophium volutator* and the prosobranch *Hydrobia ulvae Marine Biology*, **30**, 119–128.

GIERE, O. and PFANNKUCHE, O. (1982) Biology and ecology of marine oligochaeta, a review *Oceanography and Marine Biology, Annual Review*, **20**, 173–308.

GILBERT, M.A. (1973) Growth rate, longevity and maximum size of *Macoma balthica. Biological Bulletin Marine Biological Laboratory, Wood's Hole*, **145**, 119–126.

HEINLE, D.R., FLEMER, D.A. and USTACH, J.F. (1977) Contribution of tidal marshland to mid-Atlantic estuarine food-chains, in Wiley, M. (ed.) *Estuarine Processes (II)*, Academic Press, London, 309–320.

HEIP, C. and HERMAN, R. (1979) Production of *Nereis diversicolor* O.F. Müller (Polychaeta) in shallow brackish-water pond *Estuarine and Coastal Marine Science*, **8**, 297–305.

HIBBERT, C.J. (1976) Biomass and production of a bivalve community on intertidal mudflats *Journal of experimental marine biology and ecology*, **25**, 249–261.

HUMMEL, H. (1985) An energy budget for a *Macoma balthica* (Mollusca) population living on a tidal flat in the Dutch Wadden Sea *Netherlands Journal of Sea Research*, **19**, 84–92 (also 52–83).

HYLLEBERG, J. (1975) Selective feeding by *Abarenicola pacifica* with notes on *Arabenicola vagabunda* and a concept of gardening in lugworms. *Ophelia*, **14**, 113–137.

KAUTSKY, N. (1981) On the trophic role of the blue mussel *Mytilus edulis* in a Baltic coastal ecosystem, and the fate of the organic matter produced by the mussels *Kieler Meeresforsch.*, **5**, 454–461.

KRUGER, F. (1971) Bau und leben des wattwurmes *Arenicola marina Helgolander wiss. Meeresunters.*, **22**, 149–200.

KUIPERS, B.R., DE WILDE, P.A.W.J. and CREUTBERG, F. (1981) Energy flow in a tidal flat ecosystem *Marine Ecology, Progress Series*, **5**, 215–221.

JONES, N.V. (1988) Life in the Humber, invertebrate animals, in Jones, N.V. (ed.) *A dynamic estuary: man, nature and the Humber*, Hull University Press, 58–70.

LEE, W.Y. and MCALICE, B.J. (1979) Sampling variability of marine zooplankton in a tidal estuary. *Estuarine and Coastal Marine Science*, **8**, 565–582.

LOPEZ, G.R. and LEVINTON, J.S. (1987) Ecology of deposit-feeding animals in marine sediments *Quarterly review of biology*, **62**, 235–260.

MADSEN, P.B. and JENSEN, K. (1987) Population dynamics of *Macoma balthica* in the Danish Wadden Sea in an organically enriched area *Ophelia*, **27**, 197–208.

McINTYRE, A.D. and ELEFTHERIOU, A. (1968) The bottom fauna of a flatfish nursery ground *Journal of the Marine Biological Assn. U.K.*, **48**, 113–142.

McLUSKY, D.S. and ELLIOTT, M. (1981) The feeding and survival strategies of estuarine molluscs, in Jones, N.V. and Wolff, W.J. (eds) *Feeding and Survival Strategies in Estuarine Organisms*, Plenum Press, New York, 109–122.

McLUSKY, D.S., ELLIOTT, M. and WARNES, J. (1978) The impact of pollution on the intertidal fauna of the estuarine Firth of Forth, in McLusky, D.S. and Berry, A.J. (eds) *Proceedings of the 12th Europ. Mar. Biol. Symp.*, Pergamon Press, Oxford, 203–210.

McLUSKY, D.S. and McINTYRE, A.D. (1988) Characteristics of the benthic fauna, in Postma, H. and Zijlstra, J.J. (eds) *Continental Shelves*, Elsevier, Amsterdam, 131–154.

METTAM, C. (1981) Survival strategies in estuarine Nereids, in Jones, N.V. and Wolff, W.J. (eds) *Feeding and Survival Strategies in Estuarine Organisms*, Plenum Press, New York, 65–78.

MILLER, C.B. (1983) The zooplankton of estuaries, in Ketchum, B.H. (ed.) *Estuaries and Enclosed Seas*, Elsevier, Amsterdam, 103–149.

MILNE, H. and DUNNET, G. (1972) Standing crop, productivity and trophic relations of the fauna of the Ythan estuary, in Barnes, R. and Green, J. (eds) *The Estuarine Environment*, Applied Sci. Publ., London, 86–106.

MOLLER, P. and ROSENBERG, R. (1982) Production and abundance of the amphipod *Corophium volutator* on the west coast of Sweden *Netherlands Journal of Sea Research*, **16**, 127–140.

MOLLER, P. and ROSENBERG, R. (1983) Recruitment, abundance and production of *Mya arenaria* and *Cardium edule* in marine shallow waters, western Sweden *Ophelia*, **22**, 33–55.

MOORE, C.G. (1987) Meiofauna of the industrialised estuary and Firth of Forth, Scotland *Proceedings of the Royal Society of Edinburgh*, **93B**, 415–430.

MOSSMAN, D.E. (1978) *The energetics of Corophium volutator*, PhD Thesis, University of London, 153 pp.

NICHOLLS, D.J., TUBBS, C.R. and HAYNES, F.N. (1981) The effect of green algal mats on intertidal macrobenthic communities and their predators *Kieler Meeresforsch.*, Sonderh., **5**, 511–520.

NIELSEN, M.V. and KOFOED, L.H. (1982) Selective feeding and epipsammic browsing by the deposit-feeding amphipod *Corophium volutator Marine Ecology Progress Series*, **10**, 81–88.

PEER, D.L., LINKLETTER, L.E. and HICKLIN, P.W. (1986) Life history and reproductive biology of *Corophium volutator Netherlands Journal of Sea Research*, **20**, 359–373.

REVELANTE, N. and GILMARTIN, M. (1987) Seasonal cycle of the ciliated protozoan and micrometazoan biomass in a Gulf of Maine estuary *Estuarine, Coastal and Shelf Science*, **25**, 581–598.

RODDIE, B.D., LEAKEY, R.J.G. and BERRY, A.J. (1984) Salinity-temperature tolerance and osmoregulation in *Eurytemora affinis*, in relation to its distribution in the zooplankton of the upper reaches of the Forth estuary *Journal of experimental marine biology and ecology*, **79**, 191–211.

ROSENBERG, R. (1977) Benthic macrofauna dynamics, production and dispersion in an oxygen-deficient estuary of West Sweden *Journal of experimental biology and ecology*, **26**, 107–133.

SANDIFER, P.A. (1975) The role of pelagic larvae in recruitment to populations of decapod crustacea in the York River estuary and adjacent Chesapeake bay *Estuarine and Coastal Marine Science*, **3**, 269–280.

SMITH, S.L. (1978) The role of zooplankton in the nitrogen dynamics of a shallow estuary *Estuarine and Coastal Marine Science*, **7**, 555–566.

TAYLOR, C.J.L. (1987) The zooplankton of the Forth, Scotland *Proceedings of the Royal Society of Edinburgh*, **93B**, 377–388.

TUNNICLIFFE, V. and RISK, M.J. (1977) Relationships between a bivalve (*Macoma balthica*) and bacteria in intertidal sediments: Minas Basin, Bay of Fundy *Journal of Marine Research*, **35**, 499–507.

WETZEL, R.L. (1977) Carbon resources of a benthic salt-marsh invertebrate: *Nassarius obsoletus*, in Wiley, M. (ed.) *Estuarine Processes (II)*, Academic Press, London, 293–308.

WARWICK, R.M. (1971) Nematode associations in the Exe estuary. *Journal of the marine biological association, UK*, **51**, 439–454.

WARWICK, R.M. (1981) Survival strategies of meiofauna, in Jones, N.V. and Wolff, W.J. (eds) *Feeding and Survival Strategies of Estuarine Organisms*, Plenum Press, New York, 39–52.

WARWICK, R.M., JOINT, I.R. and RADFORD, P.J. (1979) Secondary production of the benthos in an estuarine environment, in Jefferies, R.L. and Davy, A.J. (eds) *Ecological processes in coastal environments*, Blackwell, Oxford, 429–450.

WILDISH, D.J. and KRISTMANSON, D.D. (1985) Control of suspension feeding bivalve production by current speed *Helgolander Meeresunters.*, **39**, 237–243.

WOLFF, W.J. (1977) A benthic food budget for the Grevelingen estuary, The Netherlands, and a consideration of the mechanisms causing high benthic secondary production in estuaries, in Coull, B.C. (ed.) *Ecology of marine benthos*, Univ. of S. Carolina Press, 297–280.

WOLFF, W.J. (1983) Estuarine benthos, in Ketchum, B.H. (ed.) *Estuaries and enclosed seas*, Elsevier, Amsterdam, 151–182.

Chapter Four

AMBROSE, W.G.Jr. (1986) Estimate of removal rate of *Nereis virens* from an intertidal mudflat by gulls (*Larus* spp.) *Marine Biology*, **90**, 243–247.

BAIRD, D. and MILNE, H. (1981) Energy flow in the Ythan estuary, Aberdeenshire, Scotland *Estuarine, Coastal and Shelf Science*, **13**, 455–472.

BERRY, A.J. (1988) Annual cycle in *Retusa obtusa* of reproduction, growth and predation upon *Hydrobia ulvae Journal of experimental marine biology and ecology*, **117**, 197–209.

BUDDEKE, R. *et al.* (1986) Food availability and predator presence in a coastal nursery area of the brown shrimp (*Crangon crangon*) *Ophelia*, **26**, 77–90.

BRYANT, D.M. (1979) Effects of prey density and site character on estuary usage by over-wintering waders *Estuarine and Coastal Marine Science*, **9**, 369–384.

BRYANT, D.M. and LENG, J. (1976) Feeding distribution and behaviour of Shelduck in relation to food supply *Wildfowl*, **27**, 20–30.

CREUTZBERG, F. *et al.* (1978) The migration of plaice larvae into the Western Wadden Sea, in McLusky, D.S. and Berry, A.J. (eds) *Proceedings 12th Europ. Mar. Biol. Symp.*, Pergamon Press, Oxford, 243–252.

EVANS, P.R. *et al.* (1979) Short-term effects of reclamation of part of Seal Sands, Teesmouth, on wintering waders and shelduck *Oecologia*, **41**, 183–206.

GEE, J.M. *et al.* (1985) Field experiments on the role of epibenthic predators in determining prey densities in an estuarine mudflat *Estuarine, Coastal and Shelf Science*, **21**, 429–448.

GOSS-CUSTARD, J.D. *et al.* (1977) The density of migratory and overwintering Redshank and Curlew in relation to the density of their prey in S.E. England *Estuarine and Coastal Marine Science*, **5**, 497–510.

GOSS-CUSTARD, J.D. and MOSER, M.E. (1988) Rates of change in the numbers of Dunlin, *Calidris alpina*, wintering in British estuaries in relation to the spread of *Spartina anglica Journal of applied ecology*, **25**, 95–109.

HAEDICH, R.L. and HAEDICH, S.O. (1974) A seasonal survey of fishes in the Mystic river, a polluted estuary in downtown Boston, Mass. *Estuarine and Coastal Marine Science*, **2**, 59–73.

HEALEY, M.C. (1970) The distribution and abundance of sand gobies, *Gobius minutus*, in the Ythan estuary *Journal of Zoology, London*, **163**, 177–229.

HOFF, J.G. and IBARA, R.M. (1977) Factors affecting the seasonal abundance, composition and diversity of fishes in a S.E. New England estuary *Estuarine and Coastal Marine Science*, **5**, 665–678.

JONES, M.B. (1976) Limiting factors in the distribution of intertidal crabs in the Avon-Heathcote estuary, Christchurch *N.Z. Journal of marine and freshwater research*, **10**, 577–587.

KREMER, P. (1977) Population dynamics and ecological energetics of a pulsed zooplankton predator, the Clenophore *Mnemiopsis leidyi*, in Wiley, M. (ed.) *Estuarine Processes (1)*, Academic Press, London, 197–205.

MEREDITH, W.H. and LOTRICH, V.A. (1979) Production dynamics of a tidal creek population of *Fundulus heteroclitus (L) Estuarine and Coastal Marine Science*, **8**, 99–118.

PIHL, L. (1982) Food intake of young cod and flounder in a shallow bay on the Swedish west coast *Netherlands Journal of Sea Research*, **15**, 419–432.

PIHL, L. (1985) Food selection and consumption of mobile epibenthic fauna in shallow marine areas *Marine Ecology, Progress Series*, **22**, 169–179.

PIHL, L. and ROSENBERG, R. (1984) Food selection and consumption of the shrimp *Crangon crangon* in some shallow marine areas in western Sweden *Marine Ecology, Progress Series*, **15**, 159–168.

PRATER, A.J. (1981) *Estuary birds of Britain and Ireland*, Poyser, T. and A.D. Calton, 440 pp.

RAFAELLI, D. and MILNE, H. (1987) An experimental investigation of the effects of shorebird and flatfish predation on estuarine invertebrates *Estuarine, Coastal and Shelf Science*, **24**, 1–13.

READING, C.J. and McGRORTY, S. (1978) Seasonal variations in the burying depth of *Macoma balthica* and its accessibility to wading birds *Estuarine and Coastal Marine Science*, **6**, 135–144.

RONN, C., BONSDORFF, E. and NELSON, W.G. (1988) Predation as a mechanism of interference within infauna in shallow brackish water soft bottoms; experiments with an infauna predator, *Nereis diversicolor* (O.F. Müller) *Journal of experimental marine biology and ecology*, **116**, 143–157.

SUMMERS, R. (1974) *Studies on the flounders of the Ythan estuary*, Ph.D. Thesis, University of Aberdeen.

THORMANN, S. (1982) Niche dynamics and resource partitioning in a fish guild inhabiting a shallow estuary on the Swedish west coast *Oikos*, **39**, 32–39.

THORMANN, S. (1986) Seasonal colonisation and effects of salinity and temperature on species richness and abundance of fish of some brackish and estuarine shallow waters in Sweden *Holoarctic Ecology*, **9**, 126–132.

THORMANN, S. and WIDERHOLM, A.-M. (1986) Food, habitat and time niches in a coastal fish species assemblage in a brackish water bay in the Bothnian Sea, Sweden *Journal of experimental marine biology and ecology*, **95**, 67–86.

VALIELA, I. *et al.* (1977) Growth, production and energy transformations in the salt-marsh killi-fish *Fundulus heteroclitus Marine Biology*, **40**, 135–144.

WIRJOATMODJO, S. and PITCHER, T.J. (1984) Flounders follow the tide to feed: evidence from ultrasonic tracking in an estuary *Estuarine, Coastal and Shelf Science*, **19**, 231–241.

WOLFF, W.J., MANDOS, M.A. and SANDEE, A.J.J. (1981) Tidal migration of plaice and flounders as a feeding strategy, in Jones, N.V. and Wolff, W.J. (eds) *Feeding and Survival Strategies in Estuarine Organisms*, Plenum Press, New York, 159–171.

ZWARTS, L. (1986) Burying depth of the benthic bivalve *Scrobicularia plana* (da Costa) in relation to siphon-cropping *Journal of experimental marine biology and ecology*, **101**, 25–39.

Chapter Five

ANDREWS, M.W. and RICKARD, D.G. (1980) Rehabilitation of the inner Thames estuary *Marine Pollution Bulletin*, **11**, 327–332.

BAYNE, B.L. (1985) (ed.) Cellular toxicology and marine pollution *Marine Pollution Bulletin*, **16**, 127–164.

BAYNE, B.L. *et al.* (1985) *The effects of stress and pollution on marine animals*, Praeger Publishers, New York, 315 pp.

BLACKSTOCK, J. (1984) Biochemical metabolic regulatory responses of marine invertebrates to natural environmental change and marine pollution *Oceanography and Marine Biology, Annual Review*, **22**, 263–313.

BOUWMAN, L.A., ROMEIJN, K. and ADMIRAAL, W. (1984) On the ecology of meiofauna in an organically polluted estuarine mudflat *Estuarine, Coastal and Shelf Science*, **19**, 633–653.

BROWN, J.R., GOWEN, R.J. and McLUSKY, D.S. (1987) The effect of salmon farming on the benthos of a Scottish sea loch *Journal of experimental marine biology and ecology*, **109**, 39–51.

BRYAN, G.W. (1984) Pollution due to heavy metals and their compounds *Marine Ecology*, **5**, 1289–1431.

BRYAN, G.W. *et al.* (1985) *A guide to the assessment of heavy metal contamination in estuaries using biological indicators*, Occasional Publication No. 4, Marine Biological Association of the UK, Plymouth, 92 pp.

BRYANT, V. *et al.* (1984) Effect of temperature and salinity on the toxicity of chromium to three estuarine invertebrates (*Corophium volutator, Macoma balthica, Nereis diversicolor*) *Marine Ecology, Progress Series*, **20**, 137–149.

CARLBERG, S.R. (1980) Oil pollution of the marine environment—with an emphasis on estuarine studies, in Olausson, E. and Cato, I. (eds) *Chemistry and Biogeochemistry of Estuaries*, J. Wiley and Sons, New York, 367–402.

CLARK, R.B. (1986) *Marine Pollution*, Clarendon Press, Oxford, 215 pp.

CONCAWE (1981) *A field guide to coastal oil spill control and clean-up techniques*, Concawe, Den Haag (Report no. 97/81), 112 pp.

COOMBS, T.L. and GEORGE, S.G. (1978) Mechanisms of immobilisation and detoxication of metals in marine organisms, in McLusky, D.S. and Berry, A.J. (eds) *Proceedings of the 12th European Marine Biology Symposium*, Pergamon Press, Oxford, 179–187.

COUGHLAN, J. (1979) Aspects of reclamation in Southampton Water, in Knights, B. and Phillips, A.J. (eds) *Estuarine and coastal land reclamation and storage*, Saxon House, London, 99–124.

DAVENPORT, J. (1982) Environmental simulation experiments on marine and estuarine animals *Advances in Marine Biology*, **19**, 133–256.

ELLIOTT, M., GRIFFITHS, A.H. and TAYLOR, C.J.L. (1988) The role of fish studies in estuarine pollution assessment *Journal of Fish Biology*, **33(A)**, 51–61.

ESSINK, K. (1978) The effects of pollution by organic waste on macrofauna in the Eastern Dutch Wadden Sea *Netherlands Institute for Sea Research, Publication Series*, **1**, 1–135.

GOWEN, R. *et al.* (1988) *Investigations into benthic enrichment, hypernutrification and eutrophication associated with mariculture in Scottish coastal waters*, Dept. of Biological Science, University of Stirling, 289 pp.

HECHT, J. (1988) America in peril from the sea *New Scientist* (9/6/88), 54–59.

HENDERSON, A.R. and HAMILTON, J.D. (1986) The status of fish populations in the Clyde estuary *Proceedings of the Royal Society of Edinburgh*, **90B**, 157–170.

HOWELL, R. (1985) The effect of bait-digging on the bioavailability of heavy metals from surficial intertidal marine sediments *Marine Pollution Bulletin*, **16**, 292–295.

HUNT, G.J. (1988) *Radioactivity in surface and coastal waters of the British Isles, 1987*, Ministry of Agriculture, Fisheries and Food, Aquatic Environment Monitoring Report, Number 19, Lowestoft, 67 pp.

JOHNSON, R. (1977) (editor) *Marine pollution*, Academic Press, London, 730 pp.

JOHNSTON, R. (1984) Oil pollution and its management *Marine Ecology*, **5**, 1433–1582.

LEATHERLAND, T.M. (1987) Radioactivity in the Forth, Scotland *Proceedings of the Royal Society of Edinburgh*, **93B**, 299–301.

LEPPAKOSKI, E. (1975) Assessment of degree of pollution on the basis of macrozoobenthos in marine and brackish water environments *Acta Academiae Aboensis (B)*, **35**, 1–89.

LOCKWOOD, A.P.M. *et al.* (1982) Microprocessor systems as an aid to estuarine studies in the laboratory *Estuarine, Coastal and Shelf Science*, **15**, 199–206.

MACKAY, D.W. (1986) Sludge dumping in the Firth of Clyde—a containment site *Marine Pollution Bulletin*, **17**, 91–95.

McLUSKY, D.S. (1982) The impact of petro-chemical effluent on the fauna of an intertidal estuarine mudflat *Estuarine, Coastal and Shelf Science*, **14**, 489–499.

McLUSKY, D.S., BRYANT, V. and CAMPBELL, R. (1986) The effects of temperature and salinity on the toxicity of heavy metals to marine and estuarine invertebrates *Oceanography and Marine Biology, Annual Review*, **24**, 481–520.

MOORE, D.C. and DAVIES, I.M. (1987) Monitoring the effects of the disposal at sea of Lothian Region sewage sludge *Proceedings of the Royal Society of Edinburgh*, **93B**, 467–477.

NELSON-SMITH, A. (1972) Effects of oil industry on shore life in estuaries *Proceedings of the Royal Society of London*, **180B**, 487–496.

NORTON, M.G. *et al.* (1981) *The field assessment of effects of dumping wastes at sea: 8, Sewage sludge dumping in the outer Thames estuary*, Fisheries Research Technical Paper, MAFF, Lowestoft, No. 62, 62 pp.

PASCOE, D. (1983) *Toxicology* Edward Arnold, London, 60 pp.

PEARSON, T.H. and ROSENBERG, R. (1978) Macrobenthic succession in relation to organic enrichment and pollution of the marine environment *Oceanography and Marine Biology: Annual Review*, **16**, 229–311.

PEARSON, T.H. and STANLEY, S.O. (1979) Comparative measurement of the Redox potential of marine sediments as a rapid means of assessing the effect of organic pollution *Marine Biology*, **53**, 371–380.

PEARSON, T.H., ANSELL, A.D. and ROBB, L. (1986) The benthos of the deeper sediments of the Firth of Clyde, with particular reference to organic enrichment *Proceedings of the Royal Society of Edinburgh*, **90B**, 329–350.

POSTMA, H. (1985) Eutrophication of Dutch coastal waters *Netherlands Journal of Zoology*, **35**, 348–359.

RIJKSWATERSTAAT (1986) *The Delta scheme*, Rijkswaterstaat, den Haag., 8 pp.

ROSENBERG, R. (1976) Benthic fauna dynamics during succession following pollution abatement in a Swedish estuary *Oikos*, **27**, 414–427.

ROSENBERG, R. (1984) (editor) *Eutrophication in marine waters surrounding Sweden—a review*, Swedish National Environmen Protection Board, Solna, Report No. 1808, 140 pp.

ROSENBERG, R. (1985) Eutrophication—the future marine coastal nuisance? *Marine Pollution Bulletin*, **16**, 227–231.

SHAW, T.L. (1980) (ed.) *An environmental appraisal of tidal power stations: with particular reference to the Severn barrage*, Pitman, London.

STEIMLE, F.W. Jr. (1985) Biomass and estimated productivity of the benthic macrofauna in the New York Bight: A stressed coastal area *Estuarine, Coastal and Shelf Science*, **21**, 539–554.

STEIMLE, F.W. and SINDERMANN, C.J. (1978) Review of oxygen depletion and associated mass mortalities of shellfish in the Middle Atlantic Bight in 1976 (*MFR Paper 1356*), *Marine Fisheries Review*, **40**, 17–26.

TALBOT, J.W. *et al.* (1982) *Field assessment of effects of dumping wastes at sea: 9, Dispersal and effects on benthos of sewage sludge dumped in the Thames estuary*, Fisheries Research Technical Paper, MAFF, Lowestoft, No. 63, 42 pp.

TENORE, K.R. *et al.* (1982) Coastal upwelling in the Rias Bajas, NW Spain: contrasting the benthic regions of the Rias de Arosa and de Muros *Journal of Marine Research*, **40**, 701–772.

TENORE, K.R., CORRAL, J. and GONZALES, N. (1985) Effects of intense mussel culture on food chain patterns and production in coastal Galicia, NW Spain *ICES C.M.*, 1985/F:62 Sess W.

THRUSH, S.F. and ROPER, D.S. (1988) Merits of macrofaunal colonization of intertidal mudflats for pollution monitoring: preliminary study *Journal of experimental marine biology and ecology*, **116**, 219–233.

VAN IMPE, J. (1985) Estuarine pollution as a probable cause of increase of estuarine birds *Marine Pollution Bulletin*, **16**, 271–276.
VAN DER VEER, H.W. *et al.* (1985) Dredging activities in the Dutch Wadden Sea: effects on macrobenthic infauna *Netherlands Journal of Sea Research*, **19**, 183–190.
WARWICK, R.M. (1986) A new method for detecting pollution effects on marine macrobenthic communities *Marine Biology*, **92**, 557–562.
WHARFE, J.R. (1977) An ecological survey of the benthic invertebrate macrofauna in the lower Medway estuary *Journal of Animal Ecology*, **46**, 93–113.
WHARFE, J.R., WILSON, S.R. and DINES, R.A. (1984) Observations of the fish populations of an east coast estuary *Marine Pollution Bulletin*, **15**, 133–136.
ZAJAC, R.N. and WHITLACH, R.B. (1982) The responses of estuarine infauna to disturbance (I & II) *Marine Ecology, Progress Series*, **10**, 1–14; 15–27.

Chapter Six

ALBERT, R.C. (1988) The historical context of water quality management for the Delaware estuary *Estuaries*, **11**, 99–107.
CARTER, R.W.G. (1988) *Coastal environments*, Academic Press, London, 617 pp.
GAMESON, A.L.H. (1982) *The quality of the Humber estuary*, Yorkshire Water Authority, Leeds, 88 pp.
HARRISON, J. and GRANT, P. (1976) *The Thames transformed*, Andre Deutsch, London, 240 pp.
MANCE, G. (1987) *Pollution threat of heavy metals in aquatic environments*, Elsevier Applied Science, London, 372 pp.
MOULDER, D.S. and WILLIAMSON, P. (1986) (editors) Estuarine and coastal pollution: Detection, Research and Control *Water Science and Technology*, **18 (4/5)**, 1–359.
SAYERS, D.R. (1986) Derivation and application of environmental quality objectives and standards to discharges to the Humber estuary (U.K.) *Water Science and Technology*, **18**, 277–285.
WILSON, J.G. (1988) *The Biology of estuarine management*, Croom Helm, London, 192 pp.
WILSON, J.G. and HALCROW, W. (1985) *Estuarine management and quality assessment*, Plenum Press, New York.

Index